FOOD SAFETY AND NUTRITION

FOOD SAFETY AND NUTRITION

JANICE R. MILLER
AND
PATRICIA E. BONETE
EDITORS

Nova Science Publishers, Inc.
New York

LIBRARY OF CONGRESS CATALOGING-IN-PUBLICATION DATA
Food safety and nutrition / Janice R. Miller and Patricia E. Bonete (editors).
p.; cm.
Includes bibliographical references and index.
ISBN 978-1-60456-335-1(hardcover)
1. Food—Safety measures—United States. I. Miller, Janice R. II. Bonete, Patricia E. [DNLM: 1. Food Industry—legislation & jurisprudence—United states. 2. Food Contamination—legislation & jurisprudence—United States. 3. Nutrition Policy—legislation & jurisprudence—United States. 4. Safety Management—legislation & jurisprudence—United states. WA 33 AA1 F686 2008]
RA601.F666 2008
363.19'26—dc22 2008008040

Published by Nova Science Publishers, Inc. ✦ New York

CONTENTS

PREFACE

Food safety is a scientific discipline describing the handling, preparation, and storage of food in ways that prevent foodborne illness. This includes a number of routines that should be followed to avoid potentially severe health hazards. Food can transmit disease from person to person as well as serve as a growth medium for bacteria that can cause food poisoning. In developed countries there are intricate standards for food preparation, whereas in lesser developed countries the main issue is simply the availability of adequate safe water, which is usually a critical item. This book presents important analyses of the issues.

Chapter 1 - Federally supported child nutrition programs/initiatives and the Special Supplemental Nutrition Program for Women, Infants, and Children (the WIC program) reach more than 39 million children and 2 million lower-income pregnant/ postpartum women. In FY2006, spending on them is anticipated to be $18.4 billion, supported by *new appropriations* of a lesser amount ($17.9 billion). The Administration's FY2007 budget request envisions spending a total of $19.1 billion, with *new appropriations* of $18.8 billion. For FY2007, the House and Senate appropriations bills (H.R. 5384) call for *spending* $19.2 billion, supported by *new appropriations* of $18.6 billion in the House and $18.9 billion in the Senate.

Child Nutrition Programs. The *School Lunch and School Breakfast programs* provide cash subsidies for all meals they serve to schools choosing to participate; larger subsidies are granted for free and reduced-price meals offered to lower-income children. The *Child and Adult Care Food program* subsidizes meals/snacks served by child care centers and day care homes; federal subsidy rules differ significantly between those provided to centers and those for day care homes. Schools and organizations operating programs for children also can receive subsidies for snacks (and, in some cases, meals) served in *after-school and other outside-of-school settings*. The *Summer Food Service program* subsidizes food service operations by public/private nonprofit sponsors volunteering to operate projects during the summer; all meals/snacks they serve are free. The *Special Milk program* operates in schools and other venues without a lunch program and subsidizes all milk served. All these subsidies are inflation-indexed and are paid only where the meals/snacks meet federal nutrition and other standards. In addition to cash aid, many providers receive *food commodities* from the Agriculture Department, at a set value per meal (and may receive "bonus" commodities from Department surplus stocks). Grants also are made to help cover *state administrative expenses*. Other significant federal programs/activities include a free fresh fruit and vegetable program in selected schools, money for a Food Service Management Institute, a small nutrition education initiative, activities to improve program integrity, meal quality, food service and safety, and support for local school "wellness policies." Separately, the WIC program provides nutrition services (e.g., nutrition education, breastfeeding support) and tailored food packages to lower-income pregnant and postpartum women, infants, and children who are judged to be at nutritional risk. And a *WIC farmers' market program* offers vouchers to WIC recipients for the purchase of fresh fruit and vegetables at farmers' markets.

These are administered by the Agriculture Department's Food and Nutrition Service and state education, health, social service, and agriculture agencies. They actually are operated, under state oversight, by over 300,000 local providers (such as schools, child care centers, health clinics). Federal payments do not necessarily cover all program costs, and nonfederal support is significant (e.g., children's families' school meal payments, state/local contributions).

Chapter 2 - In September 2006, the Food and Drug Administration (FDA) and the Centers for Disease Control and Prevention (CDC) began receiving reports on clusters of patients in various states confirmed to have *E. coli* infections. By early October, 199 people in 26 states had become ill — 102 had been hospitalized, 31 had developed hemolytic uremic syndrome (HUS), a type of kidney failure, and three had died.

Escherichia coli O157:H7 (*E. coli*) is a bacterium found in animal feces that causes diarrhea and abdominal cramps within days of exposure. An infection can lead to HUS and, in some cases, death. Public health laboratories perform a type of DNA fingerprinting to determine whether a sample taken from a patient matches those taken from other patients and contaminated food during an outbreak. The time from exposure to confirmation of an *E. coli* infection can take two to three weeks.

As the number of infections increased, an investigation that included FDA, CDC, and state and local public health officials was launched. Starting on September 14 and continuing into early October, FDA and CDC released nearly daily statements on the status of the investigation, alerting the public to the number of cases, states with confirmed cases, spinach product recalls, agency actions, and consumer advice on consumption of spinach products. Investigators were able to trace the outbreak back to several farm fields in the Salinas Valley of California. While the investigation continues, there is evidence that nearby livestock, feral pigs or other environmental sources may have contaminated one or more of the fields.

Since the outbreak, FDA has advised growers of fresh produce that they need to develop and implement voluntary guidelines to prevent outbreaks of food-borne diseases. FDA has also announced that it will convene a public meeting on the issue once the investigation is complete. Several growers groups have called for their industry to use the best agricultural and processing practices to prevent such outbreaks, not least because losses to the industry from the spinach outbreak have been estimated at $100 million.

In October 2006, the House Committee on Energy and Commerce asked FDA to explain its role in detecting contaminated food, particularly in the recent tainted spinach case. The request sought details about the agency's food safety procedures used in emergency situations.

Both proponents and opponents of the National Uniformity for Food Act (S. 3128, H.R. 4167) have weighed in on how it would have affected the *E. coli* outbreak. Opponents believe that states' ability to act would have been compromised, while proponents claim that the legislation would not have affected state adulteration or inspection authorities.

Legislators may address the recent *E. coli* outbreak during reauthorization of the farm bill in the 110th Congress, when the proposal for a single food safety agency with increased powers may be considered.

Chapter 3 - Child nutrition programs (e.g., school meal programs, summer food service, child care food programs) and the Special Supplemental Nutrition Program for Women, Infants, and Children (the WIC Program) are subject to periodic comprehensive reviews, when appropriations and other authorities expire and have to be reauthorized. They were up for reauthorization review in the 108th Congress, and the only substantial child nutrition-WIC legislation in the 108th Congress, and so far in the 109th Congress, has been the 2004 reauthorization law — the Child Nutrition and WIC Reauthorization Act of 2004, P.L. 108-265, enacted June 30, 2004.

The 2004 law extended virtually all expiring authorities through FY2009 and contained important, but incremental, changes in child nutrition programs and the WIC program; the Congressional Budget Office estimated that it will generate net new spending totaling about $230 million through FY2009. Its major feature was a set of amendments aimed at improving the integrity and administration of the school meal programs. Significant changes were made in procedures relating to the way children's eligibility for free and reduced-price school meals is certified and verified, and new initiatives to upgrade schools' administration of their meal programs were put in place. However, minimal revisions were made to the school meal programs themselves — for example, expansion of eligibility for homeless, runaway, and migrant children, loosened rules for certain higher school breakfast subsidies — and a major proposal to phase in higher income eligibility limits for free school meals was limited to an *authorization* for a pilot project. Relatively minor amendments also affected the Summer Food Service and Child and Adult Care Food programs — for example, making permanent and expanding coverage of "Lugar" rules facilitating participation by summer program sponsors and making permanent and nationally applicable a rule loosening Child and Adult Care Food program eligibility rules for for-profit child care centers.

Another area of concern addressed by the reauthorization law was nutrition, health, and nutrition education. Here, the biggest initiative was a requirement that all schools participating in school meal programs establish locally designed "wellness policies" to set nutrition, physical activity, and other goals and strategies for meeting them. Coupled with it were (1) *authorizations* for new nutrition education efforts, (2) an expansion of the program offering free fresh fruit and vegetables in selected schools, and (3) significant changes in food safety rules.

Finally, a large number of revisions were made to the law governing the WIC program. The most important among them were amendments aimed at strengthening rules that help contain food costs incurred by the program; these included provisions placing substantial limits on vendors receiving the majority of their revenue from WIC vouchers (so-called "WIC-only" stores).

Chapter 4 - For more than a century, national uniformity of food laws has been a long-sought goal of the food industry. While early statutes tried to improve uniformity, state and local requirements established in the 1800s continued in effect throughout the United States. Food traveling in interstate commerce frequently required manufacturers to meet different nutrition labeling and/or safety standards in different states, creating additional expense and confusion.

In 1990, passage of the Nutrition Labeling and Education Act (NLEA) achieved uniformity for nutrition labeling throughout the country. Provisions in the act required that all state and local provisions that were not identical to federal standards be preempted, once a study of all federal, state, and local requirements was completed. The study was conducted to determine which requirements would be preempted and which ones were unique to states/localities and should be allowed to remain in force. The NLEA allowed states to petition FDA for an exemption from preemption, if the requirement would not be in violation of any applicable federal requirement, or unduly burden interstate commerce, and provided it addressed a particular need not met by the requirements of the law. The original bill that became NLEA had contained a provision to preempt state food safety provisions, but it was dropped before passage.

Legislation has been introduced on food safety uniformity in every Congress since the 105th. In the 109th Congress, *The National Uniformity for Food Act of 2005* (H.R. 4167) would amend the Food Drug and Cosmetic Act (FDCA) to prohibit any state or other locality from establishing or continuing in effect for food traveling in interstate commerce any requirement not identical to an existing federal provision, including provisions related to adulterated food; raw agricultural commodities containing unsafe pesticides; unapproved irradiated foods; unsafe color or food additives; tolerances for poisonous ingredients; conditions for emergency permit control and their suspension; access for inspection and dietary supplement labeling regulations. The bill would allow a state to petition for an exemption or to establish a national standard related to food regulation. States would be allowed to establish requirements that otherwise would violate a FDCA provision, if the requirement is needed to address an imminent hazard that is likely to result in serious adverse health consequences. H.R. 4167 was passed by the House. *The National Uniformity for Food Act of 2006* (S. 3128) was introduced, with several changes from the House-passed version. A Senate hearing was held, but no final action was taken.

The food industry supports the bill, which it believes would improve interstate commerce by extending national uniformity to most aspects of food adulteration, preventing separate food safety notices by the 50 states and eliminating food safety warnings that are not identical with federal provisions. Opponents include food and drug officials, state attorneys general, national consumer and California groups. They believe the bill would have a serious impact on the nation's regulation of food safety and jeopardize the ability to fight bioterrorism. Opponents are also concerned about the impact of the bill on the cooperative programs for milk safety, shellfish sanitation and retail food protection. This report will be updated as action occurs.

Chapter 5 - Among many genetically engineered crops being developed are a number modified to produce pharmaceutical compounds (none yet commercialized). Proponents say such plant-based proteins could be produced more cheaply than by conventional methods and economically benefit farmers. But critics see risks, such as potential environmental, food safety, and trade problems. Among the issues is whether the current system for regulating biotechnology is adequate. This report will be updated if events warrant.

Chapter 6 - The appearance of BSE (bovine spongiform encephalopathy or "mad cow disease") in North America in 2003 raised meat safety concerns and disrupted trade for cattle and beef producers. A major issue for Congress has been how to rebuild foreign markets for U.S. beef. Other issues include whether additional measures are needed to further protect the public and cattle herd, and concerns over the relative costs and benefits of such measures for consumers, taxpayers and industry. This report will not be updated.

Chapter 7 - Media attention to food allergies is the result of the recent tracking of food allergy sufferers and a clear rise in the number of affected individuals. Several efforts are underway to improve the ability of individuals who have a food allergy to avoid products that cause symptoms that can range from mild to serious. The Food and Drug Administration (FDA) and the Food Allergy Issues Alliance each have released guidelines to address the issues of labeling and cross-contamination. Nine state attorneys general have petitioned FDA for stricter rules, which are also supported by some consumer groups. The FY2002 agriculture appropriations bill directed FDA to address and report on cross-contamination; however, this report had not yet been submitted. Most recently, the Food Allergen Labeling and Consumer Protection Act of 2004 (P.L. 108-282) was enacted on August 2, 2004. This report will provide background on food allergies and review efforts to provide improved labeling information for food allergy sufferers; it will be updated to reflect legislative or other activity.

In: Food Safety and Nutrition
Editors: J. Miller, P. Bonete, pp. 1-28
ISBN: 978-1-60456-335-1

Chapter 1

CHILD NUTRITION AND WIC PROGRAMS: BACKGROUND AND RECENT FUNDING*

Joe Richardson
Domestic Social Policy Division

ABSTRACT

Federally supported child nutrition programs/initiatives and the Special Supplemental Nutrition Program for Women, Infants, and Children (the WIC program) reach more than 39 million children and 2 million lower-income pregnant/ postpartum women. In FY2006, spending on them is anticipated to be $18.4 billion, supported by *new appropriations* of a lesser amount ($17.9 billion). The Administration's FY2007 budget request envisions spending a total of $19.1 billion, with *new appropriations* of $18.8 billion. For FY2007, the House and Senate appropriations bills (H.R. 5384) call for *spending* $19.2 billion, supported by *new appropriations* of $18.6 billion in the House and $18.9 billion in the Senate.

Child Nutrition Programs. The *School Lunch and School Breakfast programs* provide cash subsidies for all meals they serve to schools choosing to participate; larger subsidies are granted for free and reduced-price meals offered to lower-income children. The *Child and Adult Care Food program* subsidizes meals/snacks served by child care centers and day care homes; federal subsidy rules differ significantly between those provided to centers and those for day care homes. Schools and organizations operating programs for children also can receive subsidies for snacks (and, in some cases, meals)

* Excerpted from CRS Report RL33307, dated July 12, 2006.

served in *after-school and other outside-of-school settings*. The *Summer Food Service program* subsidizes food service operations by public/private nonprofit sponsors volunteering to operate projects during the summer; all meals/snacks they serve are free. The *Special Milk program* operates in schools and other venues without a lunch program and subsidizes all milk served. All these subsidies are inflation-indexed and are paid only where the meals/snacks meet federal nutrition and other standards. In addition to cash aid, many providers receive *food commodities* from the Agriculture Department, at a set value per meal (and may receive "bonus" commodities from Department surplus stocks). Grants also are made to help cover *state administrative expenses*. Other significant federal programs/activities include a free fresh fruit and vegetable program in selected schools, money for a Food Service Management Institute, a small nutrition education initiative, activities to improve program integrity, meal quality, food service and safety, and support for local school "wellness policies." Separately, the WIC program provides nutrition services (e.g., nutrition education, breastfeeding support) and tailored food packages to lower-income pregnant and postpartum women, infants, and children who are judged to be at nutritional risk. And a *WIC farmers' market program* offers vouchers to WIC recipients for the purchase of fresh fruit and vegetables at farmers' markets.

These are administered by the Agriculture Department's Food and Nutrition Service and state education, health, social service, and agriculture agencies. They actually are operated, under state oversight, by over 300,000 local providers (such as schools, child care centers, health clinics). Federal payments do not necessarily cover all program costs, and nonfederal support is significant (e.g., children's families' school meal payments, state/local contributions).

General Background

Child nutrition programs and the Special Supplemental Nutrition Program for Women, Infants, and Children (the WIC program) provide cash, commodity, and other assistance (including nutrition services and food packages in the WIC program) under three major federal laws: the National School Lunch Act (originally enacted in 1946 and renamed the Richard B. Russell National School Lunch in 1999), the Child Nutrition Act (originally enacted in 1966), and Section 32 of the act of August 24, 1935 (7 U.S.C. 612c).[1] The Agriculture Department's Food and Nutrition Service (FNS) administers the programs at the federal level; most funding is included in the annual Agriculture Department appropriations laws; and congressional jurisdiction is exercised by the Senate Agriculture, Nutrition, and Forestry Committee, the House Education and the

Workforce Committee, and, to a limited extent (relating to commodity assistance and Section 32), the House Agriculture Committee.

The most recent major amendments to the laws affecting child nutrition and WIC programs were made in the 2004 Child Nutrition and WIC Reauthorization Act (P.L. 108-265; enacted June 30, 2004) and a number of laws enacted in the 106th and 107th Congresses — most notably as part of larger measures not specifically targeted at child nutrition or WIC programs (e.g., P.L. 106-224 and P.L. 107-171). The next general reauthorization of child nutrition and WIC authorities is scheduled for 2009.

Note: For information about legislation, see (1) CRS Report 96-987, *Child Nutrition Legislation in the 104th Congress*, available on request, (2) CRS Report 97-108, *Child Nutrition Issues in the 105th Congress*, (3) CRS Report RL31578, *Child Nutrition and WIC Legislation in the 106th and 107th Congresses*, and (4) CRS Report RL33299, *Child Nutrition and WIC Legislation in the 108th and 109th Congresses*, all by Joe Richardson. For current legislative proposals in the FY2007 appropriations measures, see CRS Report RL33412, *Agriculture and Related Agencies: FY2007 Appropriations*, coordinated by Jim Monke.

Child nutrition and WIC programs are operated by a variety of local public and private nonprofit providers, and the degree of direct state involvement varies by program and state — e.g., in the WIC program, state health agencies exercise substantial control; in the school meal programs, local educational agencies (LEAs), and local "school food authorities" most often have the major role; in a very few instances, the federal government (FNS) takes the place of state agencies (for example, where a state has chosen not to operate a specific program or where there is a state prohibition on aiding private schools). At the state level, education, health, social services, and agriculture departments all have roles; at a minimum, they are responsible for approving and overseeing local providers such as schools, summer program sponsors, and child care centers, as well as making sure they receive the federal support they are due. At the local level, program benefits are provided to more than 39 million children and infants, and some 2 million lower-income pregnant and postpartum women, through just over 100,000 public and private schools and residential child care institutions, about 200,000 child care centers and family day care homes, approximately 30,000 summer program sites, and, in the case of the WIC program, some 10,000 local health care clinics/sites operated by nearly 2,000 health agencies.

All programs are available in the 50 states and the District of Columbia. Virtually all operate in Puerto Rico, Guam, and the Virgin Islands, and there are no restrictions on eligibility related to citizenship or legal residence status. American Samoa gets assistance for school lunch and WIC operations, and the

Northern Marianas receive school lunch support. In addition, WIC-like benefits are available for overseas military personnel, and Defense Department overseas dependents' schools participate in the School Lunch and Breakfast programs.

In the meal service programs like the School Lunch and School Breakfast programs, summer programs, and assistance for child care centers and day care homes, federal aid is in the form of legislatively set subsidies paid for each meal/snack served that meets federal nutrition guidelines. Most subsidies are cash payments to schools and other providers, but about 10% of the total value of assistance is in the form of federally donated food commodities. While all meals/snacks served are subsidized, those served free or at a reduced price to lower-income children are supported at higher rates. All federal meal/snack subsidy rates are indexed annually for inflation,[2] as are the income standards of eligibility for free and reduced-price meals/snacks.[3] However, federal subsidies do not necessarily cover the full cost of the meals and snacks offered by participating providers, and states and localities contribute significantly to cover program costs — as do children's families (by paying charges for nonfree or reduced-price meals and snacks). *Required* nonfederal cost-sharing (matching) is relatively minimal — states must expend at least an amount totaling just over $200 million a year nationally in order to receive federal school lunch funds. Federal per-meal/snack child nutrition subsidies may cover local providers' administrative costs, but *separate* federal payments for administrative expenses are limited to administrative expense grants to state oversight agencies, a small set-aside of funds for state audits of child care sponsors, and special administrative payments to sponsors of summer programs and family day care homes. Under the WIC program, federal appropriations pay the cost of specifically tailored food packages and include specific amounts for related nutrition services and administration.

The basic goals of the federal child nutrition programs are to improve children's nutrition, increase lower-income children's access to nutritious meals and snacks, and help support the agricultural economy. Child nutrition programs are treated as *entitlements:* federal funding and commodity support is "guaranteed" to schools and other providers based on the number of meals/snacks/half-pints of milk served, who is served (e.g., free meals/snacks to poor children get higher subsidies), and legislatively set and inflation-indexed per-meal/snack subsidy rates. On the other hand, the WIC program is a *discretionary grant* program. WIC agencies serve as many applicants as possible with the money available from federal grants (and, in some cases, voluntary state subsidies), but not necessarily all eligible applicants.

Extensive information about child nutrition programs and the WIC program may be found at the Agriculture Department's Food and Nutrition Service website: [http://www.fns.usda.gov]. This website provides information on all of the following programs — for applicants/recipients, governments, researchers, program operators, and those interested in child nutrition legislation, regulations, and guidelines — as well as recent news releases, announcements, and regulatory action.

PROGRAMS AND PARTICIPATION

School Lunch Program

Public and private nonprofit schools and residential child care institutions (RCCIs) — including Defense Department overseas dependents' schools — choosing to participate in the School Lunch program receive per-meal federal cash subsidies and federally donated commodities for *all* lunches they serve to schoolchildren. While similar aid (primarily, federally donated commodities) for school meals was provided as early as the mid-1930s, the basic School Lunch program as it operates today dates to the enactment of the 1946 National School Lunch Act and major changes to the law in the early 1960s and early 1970s.

Subsidized meals must meet federal nutrition standards based on Recommended Dietary Allowances (RDAs) and the Dietary Guidelines for Americans and certain food safety requirements. Participating schools/RCCIs also must guarantee to offer free/reduced-price meals to lower-income children, adhere to federally set administrative standards (under state oversight), and follow "Buy American" rules.[4] However, there is only minimal federal regulation of foods offered separately from subsidized school meals — "competitive foods" such as a la carte items and food sold from vending machines.

Cash subsidies are set by federal school lunch law (and indexed annually, each July, for inflation), and the amount of federal aid is not dependent on providers' costs. These cash subsidies (also called "reimbursement rates") differ depending on whether the lunch is served free, at a reduced price (no more than 40 cents), or at "full price" ("paid" meals for which a participating school or RCCI may charge as it sees fit).

Free lunch cash subsidies are paid for meals served to those who apply, and are judged eligible claiming annual family cash income below 130% of the inflation-indexed federal poverty guidelines — e.g., $21,580 for a family of three or $26,000 for a four-person family in the 2006-2007 school year. Free lunch eligibility also is extended automatically to children who are "directly certified" eligible as public assistance (e.g., food stamp) recipients, migrant and homeless children, and those served under federal runaway and homeless youth grant programs. For the 2006-2007 school year, these basic free-lunch subsidies are $2.40 a lunch.

Reduced-price lunch subsidies are paid for meals served to those who apply with family income between 130% and 185% of the inflation-indexed poverty guidelines e.g., between $21,580 and $30,710 for a family of three or between $26,000 and $37,000 for a four-person family in the 2006-2007 school year. For the 2006-2007 school year, these basic reduced-price subsidies are $2.00 a lunch.

Subsidies for *full-price ("paid") lunches* are paid for meals served to children with family income above 185% of the poverty guidelines — e.g., above $30,710 for a family of three or $37,000 for a four-person family in the 2006-2007 school year — or whose families do not apply for free or reduced-price lunches. For the 2006-2007 school year, these subsidies are 23 cents a lunch.

All of the above rates are increased by a flat 2 cents a lunch for schools/RCCIs with very high (60%+) free and reduced-price participation (about half of all lunches are subsidized with this added 2 cents).[5]

On top of cash subsidies, schools/RCCIs are entitled to federal *commodity assistance* (also discussed later in this report) for *any* lunch served. Under this rule, schools/RCCIs are entitled to receive "entitlement" commodities (valued at a minimum of 16.75 cents a lunch in the 2006-2007 school year); this amount is inflation indexed annually (each July). Two special provisions are made with regard to *fresh fruit and vegetables*: (1) $50 million taken from the amount schools are entitled to in commodities is set aside and used to make available fresh fruit and vegetables through Department of Defense distribution systems (the "DOD Fresh" program), and (2) about 400 schools in 14 states and 3 Indian reservations receive assistance to operate a program under which all students are offered free fresh fruit and vegetables.[6]

In addition to the regular School Lunch program, schools/RCCIs may expand their program to cover snacks served to children through age 18 in *after-school programs* (or other programs operating outside regular school schedules during the school year) serving an "educational enrichment purpose."[7] Federal subsidies are paid to schools operating these programs at the free snack rate offered to child care centers, if the snacks are served free to children in lower-income areas. In other cases, subsidies vary by the child's family income. (See the later discussion of the Child and Adult Care Food program for the various federal subsidy rates for snacks, as well as *separate authority* for child care organizations, including schools, to get subsidies for snacks and, in some cases, meals served free in after-school programs.)

Finally, the 2004 child nutrition reauthorization law (P.L. 108-265) requires local educational agencies participating in school meal programs to establish local *"school wellness" policies*, as they determine appropriate. These policies are to include goals/policies for nutrition education, physical activity, nutrition guidelines for all foods available on campus during the school day, and a plan for measuring implementation of the policy.

In FY2005, well more than 90% of schools and RCCIs got School Lunch program subsidies — some 95,000 schools enrolling 49.3 million children and 6,000 RCCIs with almost 300,000 children. Average daily participation in the regular lunch program during the school year was about 29.6 million children (60% of enrollment in participating schools/RCCIs). Children receiving free lunches averaged 14.5 million a day (49% of participants, or 29% of enrollment); those paying for reduced-price lunches averaged 2.9 million a day (10% of participants, or 6% of enrollment); and those buying full-price lunches averaged 12.2 million a day (41% of participants, or 25% of enrollment).[8] Average daily participation in the after-school snack component of the School Lunch program reached some 1 million children in FY2005.

School Breakfast Program[9]

As with the School Lunch program, all breakfasts meeting federal nutrition standards (and other rules applicable to the School Lunch program) are subsidized in participating public and private nonprofit schools and RCCIs, including Defense Department overseas dependents' schools. Inflation-indexed subsidy rates set by federal law vary depending on whether the breakfast is served free, at a reduced price (no more than 30 cents), or at full price. The School Breakfast program dates back to a two-year pilot project first established by the 1966 Child Nutrition Act and made permanent in 1975.

Income eligibility standards for free and reduced-price breakfasts are the same as in the School Lunch program (see the earlier discussion), and, for the 2006-2007 school year, basic cash subsidies are $1.31 per *free breakfast*, $1.01 per *reduced-price breakfast*, and 24 cents per *full-price breakfast*.[10] Special "severe need" rates (an extra 25 cents for each free or reduced-price breakfast, indexed annually) are paid to schools and RCCIs with relatively high (40%+) free and reduced-price participation, and the majority of breakfasts are subsidized at this higher rate. With the exception of different subsidy rates and the lack of a specific entitlement to commodity support, the School Breakfast program operates very much like the School Lunch program, although in fewer schools and with a lower rate of participation among enrolled children. As with the School Lunch program, participation in the School Breakfast program by schools and RCCIs is voluntary — although a number of states have enacted laws requiring some schools with lunch programs to join the breakfast program.

In FY2005, 80% of School Lunch program schools and virtually all RCCIs in the lunch program also operated a breakfast program — i.e., some 76,000 schools enrolling 40.8 million children and about 6,000 RCCIs enrolling almost 300,000 children. Average daily participation in the breakfast program during the school year was 9.4 million schoolchildren (about 23% of enrollment). Children receiving free breakfasts formed the bulk of participants, averaging 6.8 million a day (72% of participants, or 17% of enrollment); those getting reduced-price breakfasts averaged 900,000 a day (10% of participants, or 2% of enrollment); and those buying full-price breakfasts averaged 1.7 million a day (18% of participants, or 4% of enrollment).

Child and Adult Care Food Program (CACFP)[11]

The CACFP dates to 1968, when federal assistance for programs serving children outside of school ("special food service" programs) was first authorized. In 1975, the summer food service and child care components were first formally separated as individual programs.

Public, private nonprofit, and (under certain conditions) for-profit nonresidential *child care centers* — typically serving 30-50 children or more — choosing to participate in the CACFP receive cash subsidies for each meal or snack they serve (up to two meals and one snack, or two snacks and one meal per child a day, or three meals a day in emergency homeless shelters). Eligible centers also include after-school projects and Head Start centers, as well as recreational centers, centers caring for children with disabilities, and residential emergency homeless shelters. In order to qualify for subsidies, meals/snacks must meet federal nutrition standards and be served to children age 12 or under (or migrant children age 15 or under, or children with disabilities); in the case of emergency/homeless shelters, the age limit is 18. In addition, participating centers receive commodity assistance (see the later discussion of commodity distribution) or cash-in-lieu of commodities; this represents about 4% of the value of aid they get.

Inflation-indexed federal cash subsidies to centers vary by the type of meal served (breakfast, lunch/supper, snack). Similar to the school meal programs, these subsidies vary by whether they are served to (1) children with family income below 130% of the federal poverty income guidelines (for those who would be eligible for a free school meal, see the School Lunch program discussion), (2) children with family income between 130% and 185% of the poverty guidelines (those who would be eligible for a reduced-price school meal, see the School Lunch program discussion), or (3) children with family income above 185% of the poverty guidelines (those who would not be eligible for either a free or reduced-price school meal).[12]

Subsidies for lunches (or suppers) and breakfasts are the same as those noted above for the School Lunch and Breakfast programs. For July 2006-June 2007, the subsidies for snacks are 65 cents for "free" snacks, 32 cents for "reduced-price" snacks, and 6 cents for all other snacks. However, unlike the school meal programs, while federal cash subsidies differ according to the family income of individual children in a center, there is no requirement that free or reduced-price meals be served. Centers may adjust their fees to account for the federal subsidies or charge (or not charge) separately for meals to account for the subsidies; but the CACFP itself does not regulate the fees they charge.

The CACFP generally operates through *child care centers* that are public or private nonprofit entities. However, *for-profit* child care centers can participate in the CACFP if they meet certain conditions. They are qualified if they receive at least some payments derived from Title XX of the Social Security Act (the federal Social Services Block Grant) for at least 25% of enrolled children.[13] In addition, under a pilot project that began operating in Iowa and Kentucky, a more liberal test can be applied to for-profit centers: they may participate if at least 25% of enrolled children meet the family income requirements for free/reduced-price school meals; this pilot was expanded to Delaware in FY2002. Finally, under provisions of law enacted in December 2000 (P.L. 106-554), the more liberal Iowa/Kentucky/Delaware rule was made applicable *nationwide*. The nationwide authority granted in the December 2000 law originally covered FY2001 only. But a series of laws — P.L. 107-76, P.L. 108-134, and P.L. 108-211 — extended it through June 30, 2004, and P.L. 108-265 made it *permanent and nationally applicable*.[14] As a result, there now are three potential methods by which for-profit centers can qualify: the original Title XX rule, the "pooling" variation of that rule (see footnote 13), and, most important, the (now permanent and nationwide) Iowa/Kentucky/Delaware rule.

In addition to the regular CACFP, the law allows organizations that would normally be eligible to participate in the regular CACFP as child care centers (including schools), and running *after-school programs* (or other similar programs operated outside regular school schedules) serving an "educational enrichment purpose," to get federal CACFP subsidies for snacks served free in their programs to children (*through age 18*) in lower-income areas — at the free snack rate noted above.[15] Moreover, in seven states — Delaware, Illinois, Michigan, Missouri, Pennsylvania, New York, and Oregon — the law allows federal subsidies to be offered for free *meals,* typically suppers, served in after-school programs (at the free lunch rate, $2.40 a supper).

Separately, the CACFP provides cash subsidies to family and group *day care homes*, typically serving four to six children. This component operates very differently than the component for centers, although, like centers, subsidies for day care homes are limited to two meals and one snack (or one meal and two snacks) per day per child.

Day care homes receive cash subsidies that generally do *not* differ by individual children's family income — unlike the subsidy structure in programs for schools and child care centers, which differs according to the family income of the child to whom the meal/snack is served.[16] Instead, there are two distinct sets of subsidy rates that generally depend on the location of the home or the provider's income. "Tier I" homes — those located in lower-income areas or operated by lower-income providers — receive higher cash subsidies; for July 2006-June 2007, all lunches/suppers are subsidized at $1.97 each, all breakfasts are subsidized at $1.06, and all snacks are subsidized at 58 cents. The majority of participating homes are in Tier I. On the other hand, "Tier II" homes — those not located in lower-income areas or without a lower-income provider — receive much lower subsidies; for July 2006-June 2007, all lunches/suppers are subsidized at $1.19 each, all breakfasts at 39 cents, and all snacks at 16 cents. Tier II homes may seek the higher Tier I rates for individual lower-income children, and, similar to the program in centers, day care home sponsors may opt to have subsidies calculated according to the family income demographics of the children in the homes they sponsor — if family income documentation is obtained.[17]

Day care homes participate under the aegis of public or private nonprofit "sponsoring organizations" that handle administrative tasks (e.g., overseeing compliance with program requirements, making federal subsidy claims). These sponsors receive separate inflation-indexed monthly payments for their administrative/oversight costs, varying according to the number of homes the sponsor oversees; for July 2006-June 2007, these per-home payments ranged from $50 to $95 a month. Centers may participate either directly as independent centers or through a sponsoring organization; but center sponsors do not receive additional federal administrative funds (although sponsors can assess centers for administration to a limited degree). Participating day care homes and centers generally must meet state or local licensing or other state-set approval requirements (or certain alternate federal standards if there are no state or local rules applying to them).

Finally, the CACFP funds state costs connected with auditing sponsors and providers. States are provided an annual amount equal to 1% of their CACFP subsidies, and unused money can be reallocated among states that need it.

In FY2005, some 47,000 centers/sites (with some 19,000 sponsors) with an average daily attendance of 2.1 million children participated — 31% (660,000) of the children were in for-profit centers/sites; 5% (115,000) participated in outside-of-school-hour centers/sites; and 26% (540,000) were served in Head Start centers/sites. In addition, 153,000 day care home sites (with about 1,000 sponsors) received subsidies for an average daily attendance of 900,000 children.

Summer Food Service Program[18]

Local public and private nonprofit "service institutions" running youth/recreation programs, summer feeding projects, or camps can receive cash subsidies and some federally donated food commodities for free food service to children aged 18 and younger (and older disabled children) during the summer. Participating service institutions (also called sponsors) often, but not of necessity, are entities that provide ongoing year-round service to the community, and include schools, local government agencies, camps, colleges and universities in the National Youth Sports program, and (with some restrictions governing the number of sites and children served) private nonprofit organizations (e.g., churches).

Summer programs date to 1968, when federal assistance for "special food service" programs serving children outside of school was first authorized. In 1975, the summer and child care food service components of the old special food service program were first formally separated as individual programs.

Sponsors of three types of summer program sites can be approved: (1) "open" sites operating in lower-income areas where 50% or more of the children have family income below 185% of the federal poverty guidelines (i.e., more than half of the children are eligible for free or reduced-price school meals), (2) "enrolled" sites where at least half of the children enrolled in a sponsor's program (e.g., a summer education or recreation activity) are eligible for free or reduced-price school meals, and (3) summer camps. Summer meals/snacks are provided free to all children at open or enrolled sites and to lower-income children in camps.

Summer sponsors get operating cost subsidies for all meals/snacks served free — one meal and one snack, or two meals (three meals for children in programs for migrant children) per child per day. These subsidies cover documented food and food service costs up to annually indexed (each January) per-meal/snack maximums.[19]

For the summer of 2006, the maximum operating cost subsidy rates are $2.56 for each lunch/supper, $1.47 for breakfasts, and 59 cents for snacks. Subsidies do not vary by individual children's family income, and most sponsors receive the maximum allowable rates. Summer program sponsors also receive significant payments for administrative costs (up to about 27 cents a meal) according to the number and type of meals/snacks served and the type of program (e.g., urban vs. rural sites, self-preparation vs. contracted vendor preparation), and state agencies receive special administrative cost payments for oversight (see later discussion of state administrative expenses) and health inspections (see below). Schools wishing to sponsor summer programs may effectively bypass most separate Summer Food Service program rules and, under a "seamless summer option," operate a summer program using School Lunch and School Breakfast program rules and subsidy rates.

In addition to subsidies to sponsors, states receive direct subsidies for health and meal quality inspection costs related to summer programs — an amount equal to 1% of a state's summer program subsidies.

In July 2005, nearly 3,800 sponsors operating some 30,000 sites provided subsidized meals and snacks to an average daily attendance of just under 2 million children. In addition, at least 1.6 million children received summer meals subsidized through the *School Lunch program* (1.4 million of these children received free or reduced-price meals).

Special Milk Program

Under this program, schools and institutions like summer camps and residential and non-residential child care facilities not otherwise participating in a federally subsidized meal service program, along with schools with split (part-day) sessions for kindergartners or pre-kindergartners where the children do not have access to regular school meal programs, provide milk to all children at a reduced price or free. Each half-pint served is federally subsidized at a different rate, depending on whether it is served free (to those meeting criteria for free school lunches) or not — but the provision of free milk to needy children is up to the participating school and is not required. Half-pints are subsidized at 14.5 cents a half-pint for the 2006-2007 school year, or at their net *cost* (typically 1-2 cents higher) if served free. Participating schools and other outlets must have a policy of lowering any prices charged for milk they serve to the maximum extent possible, and of using their federal payments to reduce the price of milk to children. Although similar assistance existed in prior years, this program dates to 1954-1955.

In FY2005, almost 7,000 schools and other outlets served about 100 million subsidized half-pints (7% free) to roughly 500,000 children.

Commodity Assistance[20]

The Agriculture Department provides commodity support for School Lunch program schools, the CACFP, and the Summer Food Service program. Federal donations of food commodities for child nutrition operations began in the mid-1930s to support the agricultural economy (most prominently, following the enactment of Section 32 of the Agricultural Act of August 24, 1935). In 1974, a basic "entitlement" to commodity assistance was placed in child nutrition law.

In addition to cash subsidies, schools (which receive the bulk of federally donated commodities) and other providers are "entitled" to a specific dollar value of commodities based on the number of meals they serve.[21] The inflation-indexed (each July) commodity entitlement is set at 16.75 cents a meal for the 2006-2007 school year. The Department purchases these commodities and pays for most processing costs to fulfill this guarantee, with the goals of meeting the preferences of recipient agencies, supporting agricultural prices, and removing agricultural surpluses.

Schools and other providers also receive "bonus" commodities donated from federal stocks acquired, at the Department's discretion, for agricultural support reasons (e.g., surplus commodities and excess Commodity Credit Corporation holdings). These bonus commodities were valued at $120 million in FY2005.

Finally, the value of commodity donations (both "entitlement" and "bonus") must equal at least 12% of the total cash and commodity assistance provided under the School Lunch program. If this 12% requirement is not met, the Agriculture Department is required to purchase and distribute additional commodities to schools; because of the value of bonus commodity donations, this 12% requirement has not triggered added commodity purchases/distributions in recent years.

State Administrative Expenses

Under authority in the Child Nutrition Act tracing back to 1966, states receive grants to help cover general administrative and oversight/monitoring costs associated with child nutrition programs (including commodity distribution costs, but not including WIC program costs). The national amount each year is equal to 1.5% of federal cash payments for the School Lunch, School Breakfast, Special Milk, and Child and Adult Care Food programs. The majority of this money is allocated to states based on their share of spending on the programs covered above; about 15% is allocated under a "discretionary" formula granting each state additional amounts for the Child and Adult Care Food program, commodity distribution, and "coordinated review efforts" (see the later discussion of other child nutrition special projects and activities for a description of coordinated review efforts). In addition, states receive administrative payments for their role in overseeing summer programs — equal to approximately 2.5% of their summer program aid. States are free to apportion their various federal administrative expense payments among child nutrition initiatives (including the summer program) as they see fit.

The WIC and WIC Farmers' Market Programs[22]

The Special Supplemental Nutrition Program for Women, Infants, and Children (the WIC program) provides nutritious foods and other support to lower-income pregnant, postpartum, and breastfeeding women, and to infants and children (up to age 5).[23] The program is operated through some 10,000 local health care clinics/sites run by 2,000 local agencies and state health agencies (and more than 30 Indian tribal organizations participating as separate grantees treated like states). In addition, a special WIC-like extension run through the Defense Department serves overseas military personnel.[24] The WIC program originated as a two-year pilot project in 1972, and was established in law as a national program in 1975.

Although the administering state and tribal WIC agencies have some discretion in judging eligibility, recipients' household income generally can be no higher than 185% of the federal poverty guidelines (the same standard used for determining eligibility for reduced-price school meals) — e.g., $24,440 for a family of two or $30,710 for a three-person family, until the next annual inflation adjustment in July 2007 — and all state agencies now use the 185% income test. In the alternative, applicants who are recipients of Medicaid, food stamp, or

Temporary Assistance for Needy Families (TANF) benefits may be judged automatically income-eligible. However, just as important as meeting the income test, enrollees also must be judged at "nutritional risk" by health professionals in the health agencies and clinics that administer the program — e.g., based on clinical measurements, documented nutritionally related medical conditions, dietary deficiencies.

Foods are provided ("prescribed") through vouchers/checks, listing the specific foods and amounts appropriate to the recipient's status, that are redeemed at approved retail outlets (or, much less commonly, supplied directly by the administering agency itself).[25] The specific foods prescribed (e.g., juice, infant formula, cereal, eggs) are based on a set of federally established food packages that differ by recipient type (e.g., infant, pregnant mother). However, WIC agencies have considerable leeway in implementing the federally defined food packages. They choose *which* infant formulas (or other items like juices or cereals) are offered to meet the federal food package requirements and how to respond to recipients' special needs.

Participating retailers (more than 45,000 are approved) then redeem the vouchers/checks for cash through arrangements with their state WIC agency. The program also provides financial support for state and local clinic Nutrition Services and Administrative (NSA) costs — about 27% of total federal aid provided to states and tribal organizations. These include costs associated with nutritional risk, health, and immunization assessments, nutrition and substance abuse education and counseling, health care and immunization referrals, certain breastfeeding promotion and support costs, determining eligibility, and issuing and redeeming vouchers (or directly delivering food items or operating an EBT system).[26] In addition, about 1% of WIC spending goes for grants to improve infrastructure (e.g., computer systems), technical assistance, and special projects like breastfeeding peer counselors and help with state management information systems.

Annual federal appropriations are granted to state and Indian tribal WIC agencies under a formula that reflects food and NSA caseload costs, inflation, and "need" (as evidenced by poverty indices) — although small amounts are set aside and distributed at the Agriculture Department's discretion for infrastructure development (like building and upgrading EBT and management information systems) and other (e.g., breastfeeding counselor) projects, and other funds have been used for small special initiatives (e.g., immunization and health care outreach efforts). These annual new appropriations are supplemented by unused money carried over from year to year and reallocated among state and tribal grantees or retained by state WIC agencies for use in the next year. Moreover, recent appropriations also have included a small ($125 million) contingency fund.

The direct support supplied by appropriations and funding carryovers is added to by the effect (holding down food costs) of substantial rebates (discounts) from manufacturers with which state agencies contract (through competitive bidding) to be the sole source for a particular food item. This cost control strategy is used by all state agencies for infant formula, and, in FY2005, the majority of them received 90%-97% discounts on thc wholesale cost of infant formula.[27] Moreover, 11 state agencies also have rebate contracts for juice and cereal. In FY2005, the estimated rebate savings was worth $1.6 billion.[28] As part of this cost containment effort, special limits are in place for approving retailers who derive the majority of their revenue from WIC vouchers (so-called "WIC-only" stores); at present, no new WIC-only stores may be certified for participation, and the prices they may charge the WIC program are significantly constricted.

Finally, a relatively small (in dollar terms) *WIC farmers' market nutrition program* operates, with significant non-federal matching funding (30% of administrative costs), in 38 states, the District of Columbia, Puerto Rico, Guam, and four Indian tribal organizations. It is run by a variety of state agencies (typically, state agriculture offices) coordinating with the regular WIC program and offers some 2.5 million WIC participants special vouchers that are used to buy fresh produce through about 4,700 participating farmers' markets, roadside stands, and community supported agriculture projects. The federal government pays the cost of the first $30 (per participant per year) of each farmers' market voucher, and shares administrative costs with the states and other political jurisdictions sponsoring the program.

In FY2005, average monthly WIC participation was 8.0 million persons — 1.97 million women, 2.05 million infants, and 4.0 million children. Average per-person federal costs were $37.50 a month for food and $14.40 a month for NSA expenses.

Special Projects and Other Support Activities[29]

Under the *coordinated review effort (CRE)*, the FNS, in cooperation with state agencies, conducts periodic school compliance and accountability evaluations to improve management and identify administrative, subsidy claim, and meal quality problems. This $5 million-a-year effort is the major ongoing initiative related to maintaining the integrity of child nutrition programs (although, under the terms of the most recent child nutrition reauthorization law,

P.L. 108-265, new funding is available for additional administrative reviews of schools needing special oversight). The Agriculture Department's Economic Research Service (ERS) and the FNS conduct *nutrition research, studies, surveys, and evaluations* (typically totaling $6-$8 million a year for child nutrition and WIC activities); this work was formerly done exclusively by the FNS.[30] A national *Food Service Management Institute (FSMI)* provides technical assistance, instruction, materials in nutrition and food service management (it is funded at $4 million a year). And an *information clearinghouse* provides information to support community-sponsored food assistance initiatives (funding of about $250,000 a year). *Special FNS projects* — "Team Nutrition" nutrition education projects, a food safety project, technical assistance to program operators, food service training grants, aid with electronic food service resource systems, "program integrity" initiatives — are aimed at helping schools and other providers with nutrition education materials, assisting them to improve their meal service operations and the quality of meals, and ensuring federal support is spent correctly; they are typically funded at about $10-$20 million a year.

In addition to the above-noted ongoing projects and activities, the 2004 Child Nutrition and WIC Reauthorization Act (P.L. 108-265) added new, mostly one-time, "mandatory" spending to assist states and schools in *improving administration of their school meal programs* (especially determination of eligibility for free and reduced-price meals), gave money to the FNS to help local schools develop *"wellness policies"* (outlined in the school lunch section of this report), and provided one-time funding for specific *pilot projects* (related to expanding summer food service). It also extended, expanded, and provided new money (later increased) for a *free fresh fruit and vegetable program* in selected schools (discussed under the description of the School Lunch program).[31]

FUNDING FOR CHILD NUTRITION AND WIC PROGRAMS

Federal support for child nutrition and WIC programs is derived from funding provided out of (1) annual Agriculture Department appropriations, (2) permanent appropriations not included in the annual appropriations laws (e.g., money directly appropriated for the Food Service Management Institute under its authorizing law and other spending mandated by law), (3) unused money available (carried over) from prior years' appropriations, transferred from other Agriculture Department appropriations accounts, or recovered from states and operators, and (4) funds paid for child nutrition initiatives from budget accounts separate from appropriations to the child nutrition and WIC accounts (e.g., a large share of commodity assistance).

Actual spending for most child nutrition programs — but not WIC programs —normally is dictated by the demand for federal dollars called for by the number and type of subsidized meals, snacks, or half-pints of milk served, not the funding made available (annually appropriated or otherwise); these programs are mandatory "entitlements." WIC (and WIC farmers' market) spending, on the other hand, is "discretionary" and dictated by the dollar amounts available from current and prior-year (carried-over or recovered unused) appropriations. State WIC agencies serve as many persons as they can within their respective allocations of appropriated funds (supplemented by carryovers and recoveries), and not all eligibles may be served if funds run short. Individual programs within the child nutrition budget account (e.g., the School Lunch and Breakfast programs) do not receive specific ("line-item") appropriations, and thus funding may be shifted among the various child nutrition programs as needed — so long as total spending stays within the overall amount available from new appropriations and other sources.

As a result, readily identifiable annual congressional *appropriations* — typically divided into two major accounts, a *child nutrition account* and a *WIC program account* — do not provide a clear or complete picture of total federal support for child nutrition and WIC programs in a given year.[32] Rather, *spending* figures —normally higher than appropriations, shown in this report's second table (typically, obligations), give a much better overview than appropriations amounts.

Table 1 presents recent *appropriations* for the child nutrition and WIC program accounts (FY2004-FY2006), the appropriations request for FY2007, the House-approved FY2007 appropriations bill (H.R. 5384; H.Rept. 109-463), and the Senate-reported FY2007 appropriations measure (H.R. 5384; S.Rept. 109-266). Table 2, on the other hand, shows, for the same years, *spending* figures (estimates for FY2006 and FY2007) for the various child nutrition programs and activities, the WIC program, and the WIC Farmers' Market program.

Table 1. Child Nutrition and WIC Appropriations: FY2004-FY2007 (in millions)

Annual appropriations account	FY2004: Consolidated Appropriations (P.L. 108-199)	FY2005: Consolidated Appropriations (P.L. 108-447)	FY2006: Agriculture Appropriations (P.L. 109-97)	FY2007: Administration request	FY2007: House bill (H.R. 5384)	FY2007: Senate bill (H.R. 5384)
Child nutrition	$ 11,417.4	$ 11,782.0	$ 12,660.8	$ 13,645.5	$ 13,345.5	$ 13,654.5
WIC	4,611.9	5,277.2	5,204.7	5,200.0	5,244.0	5,264.0
Total	$ 16,029.3	$ 17,059.2	$ 17,865.5	$ 18,845.5	$ 18,589.5	$ 18,918.5

Sources: Agriculture Department appropriations laws for FY2004-FY2006, annual Agriculture Department budget documents, H.Rept. 109-463, and S.Rept. 106-266.

Notes: WIC appropriations have been reduced to reflect directed rescissions in discretionary programs — $27.3 million (FY2004), $42.2 million (FY2005), and $52.3 million (FY2006) — and a $100 million transfer in FY2004. *Not* included are mandatory appropriations made by laws outside the appropriations acts, funding available from other Agriculture Department accounts (e.g., some commodity support, a $120 million transfer from the Food Stamp budget account in FY2005), unused carryover or recovered funding from previous years, and money for the WIC farmers' market program in FY2005 and later years (about $20 million a year appropriated through another budget account). The House bill's child nutrition appropriation does not include a $300 million contingency fund proposed by the Administration and included in the Senate bill; however, it (like the Senate bill and the Administration's request) does allow for $125 million in contingency funds for the WIC program. Added funds for a free fresh fruit and vegetable program in selected schools are not included in the House bill amounts shown here (they are appropriated separately), but are included in the Senate child nutrition appropriation.

Table 2. Continued

Table 2. Child Nutrition and WIC Spending: FY2004-FY2007 (in millions)

Programs/Activities	FY2004 (actual)	FY2005 (actual)	FY2006 (estimated)	FY2007: (estimated, Administration request)	FY2007: (estimated, House bill, H.R. 5384)	FY2007: (estimated, Senate bill, H.R. 5384)
School lunch a	$ 6,629.7	$ 7,023.8	$ 7,457.6	$ 7,831.9	$ 7,831.9	$7, 831.9
School breakfast a	1,786.9	1,937.4	2,082.9	2,251.3	2,251.3	2,251.3
Child & adult care food a	2,055.7	2,134.4	2,156.4	2,272.1	2,272.1	2,272.1
Summer food service [a]	266.6	273.2	290.2	305.9	305.9	305.9
Special milk	14.0	16.6	15.0	14.0	14.0	14.0
Commodity support [b]	878.0	941.6	922.7	959.6	959.6	959.6
State admin. expenses	139.6	145.7	156.1	165.5	165.5	165.5
Coordinated review effort	5.2	5.2	5.2	5.3	5.3	5.3
Food service management institute (FSMI)	3.0	4.0	4.0	4.0	4.0	4.0
Fresh fruit and vegetable program in schools [c]	3.0	9.0	15.0	9.0	25.0	18.0
Special projects & other activities [d]	16.2	18.6	35.6	18.2	18.2	18.2
Child nutrition total	11,797.9	12,509.5	13,140.7	13,836.8	13,852.8	13,845.8
WIC program [e]	4,899.4	5,002.8	5,237.8	5,225.6	5,354.0	5,374.0

Table 2. Continued

Programs/Activities	FY2004 (actual)	FY2005 (actual)	FY2006 (estimated)	FY2007: (estimated, Administration request)	FY2007: (estimated, House bill, H.R. 5384)	FY2007: (estimated, Senate bill, H.R. 5384)
WIC farmers' markets [f]	20.9	24.8	19.8	19.8	20.0	20.0
Overall total	$ 16,718.2	$ 17,537.1	$ 18,398.3	$ 19,082.2	19,226.8	19,239.8

Sources: Agriculture Department budget documents, H.Rept. 109-463, S.Rept. 109-266, and estimates from the Agriculture Department's Food and Nutrition Service for WIC spending under the House bill.

Notes: Includes spending from all funding sources: regular appropriations, money from other budget accounts (e.g., Section 32 commodity support, a $120 million FY2005 transfer from the Food Stamp account), permanent appropriations, and carryovers of unused funding from previous fiscal years. Does *not* include spending on child nutrition- and WIC-related research.

[a] Figures include cash assistance only, not the value of commodities or cash-in-lieu of commodities.

[b]. Figures include (1) some $400 million in commodities purchased for and donated to child nutrition programs (at no charge to the child nutrition account) to meet commodity entitlements, (2) cash subsidies and commodity letters of credit provided in lieu of commodity entitlements (e.g., $94 million in FY2005), and (3) certain federal commodity donation administrative/distribution/computer support spending (e.g., $9 million in FY2005). *Not* shown is the value of "bonus" commodities donated by the Department from excess federal commodity holdings (e.g., $120 million in FY2005).

[c] Unlike other numbers in this table, figures for the free fresh fruit and vegetable program in schools represent funding made *available* for each year, not actual spending, because of uncertainty about spending amounts. Actual spending figures are likely to be substantially lower than the amounts shown. The table allocates to FY2004 $3 million of the $6 million provided by the 2002 farm bill (P.L. 107-171) for FY2003 and FY2004. The $9 million shown for FY2005 is the amount mandated by the 2004 child nutrition reauthorization law (P.L. 108-265) for each year beginning with FY2005. The $15 million shown for FY2006 includes the $9 million mandate from the child nutrition reauthorization law, plus $6 million appropriated by the FY2006 appropriations measure. The Administration's request for FY2007 includes only the $9 million mandated amount. The House $25 million figure is a discretionary appropriation that assumes suspension of the $9-million-a-year mandate. The Senate figure includes the $9 million mandatory funding amount, plus a $9 million discretionary appropriation.

[d] Special projects and other activities include nutrition education efforts through "Team Nutrition," food safety education, several program integrity/technical assistance initiatives undertaken by the Administration or directed by the 2004 child nutrition

Table 2. Continued

reauthorization law, an information clearinghouse, program evaluations, a summer program pilot project, and funding for the FNS to help schools establish "wellness policies." A large portion of the FY2006 amount reflects one-time spending directives in the 2004 child nutrition reauthorization law (P.L. 108-265).

e. The FY2007 Administration request for the WIC program is predicated on enactment of a recommendation to limit WIC Nutrition Services and Administration (NSA) support to 25% of state grants (an estimated savings of $152 million in FY2007). The House and Senate figures reject the proposed NSA limits. They also reflect major changes in assumptions as to food costs, participation, and the availability of prior-year funds (as estimated by the Food and Nutrition Service after passage of the House bill and before reporting of the Senate bill).

e. WIC farmers' market figures represent *spending* from amounts available from new annual appropriations, *plus* money recovered or carried over from prior years. They reflect a substantial amount of unused funding in FY2004 (later made available and obligated in FY2005), and they assume significant, although shrinking, recoveries/carryovers (a decrease in unused money) in FY2005, FY2006, and FY2007. Annual *appropriations* were $22.9 million in FY2004, $19.8 million in FY2005, and $19.8 million in FY2006. A $19.8 million appropriation is requested for FY2007. In effect, the drop in appropriations beginning with FY2005 does not show in spending totals until FY2006 because unspent funds from earlier years were available. However, the FY2005 spending figure may be revised downward by as much as $2 million when all unused money is recovered, and the FY2006 and FY2007 amounts may be revised upward to the extent recoveries from FY2005 are made available as grants. The FY2007 figure for the House and Senate bills reflects $200,000 added by these bills.

REFERENCES

[1] The School Lunch and Child Nutrition Acts provide most of the basic authorities for child nutrition and WIC programs. For information on what each law covers, see CRS Report RL33299, *Child Nutrition and WIC Legislation in the 108th and 109th Congresses,* by Joe Richardson. Section 32 authority provides funding for cash child nutrition subsidies (permanent appropriations under Section 32 are transferred to the child nutrition account annually) and the acquisition of food commodities for distribution to child nutrition programs (Section 32 money is used to buy surplus and distribute commodities). For more information, see CRS Report RS20235, *Farm and Food Support under USDA's Section 32 Program*, by Geoffrey Becker.

[2] Most often, using the "food away from home" component of the Consumer Price Index (CPI) for All Urban Consumers.

[3] Cash subsidy rates and income eligibility standards typically differ (are higher) for Alaska and Hawaii. However, (1) while free milk eligibility standards vary for Alaska and Hawaii in the Special Milk program, federal subsidies do not; and (2) since commodity support is provided without regard to the free/reduced-price character of meals served, commodity assistance rates do not differ for Alaska and Hawaii. Cash subsidy rates and eligibility standards for the District of Columbia, Guam, Puerto Rico, and the Virgin Islands are those for the contiguous 48 states.

[4] "Buy American" rules require that participating schools purchase U.S.-produced agricultural commodities and food products processed in the U.S. using U.S.-produced commodities — *to the maximum extent practicable.* These rules apply to schools located in the continental U.S. In addition, a special rule directs schools in Hawaii and Puerto Rico to buy commodities or food products produced there, if they are produced in sufficient quantities to meet schools' needs.

[5] Participating schools may offer *all* meals free and not make annual free/reduced-price eligibility determinations for individual students, or separately count free, reduced-price, and full-price meals — if they pay any extra cost (i.e., claim from the federal government only the estimated amount they would have received if they had operated a regular free, reduced-price, and full-price meal program). This choice generally is used by schools with very high proportions of needy children. It reduces the burden of making individual eligibility determinations and simplifies daily meal counts and procedures for claiming federal subsidies, thereby saving

schools administrative costs. Three options to accomplish this are offered to schools. The two most widely used are named "provision 2" and "provision 3." Under *provision 2*, schools make free/reduced-price eligibility determinations in the first year of a four-year cycle; in the following three years, they count the total number of meals served, and the percentages of free, reduced-price, and full-price meals served in the first year are applied to the total meal count to calculate their federal subsidies. Under *provision 3*, schools can, for four years, receive federal subsidies equal to those received in the last year in which they made free/reduced-price eligibility determinations, adjusted for enrollment changes and inflation. For both provisions, schools may be approved for four-year extensions if the composition of their school population remains stable.

[6] This free fresh fruit and vegetable program is funded separately from the School Lunch program. It began as a pilot project established by the 2002 "farm bill" (P.L. 107-171), and was expanded, given mandatory annual funding, and made permanent by the 2004 child nutrition reauthorization law (P.L. 108-265); $9 million a year was provided to operate the program in selected schools in eight states and three Indian reservations. Later law (P.L. 109-97) expanded the number of covered states (to 14) and added funding ($6 million for FY2006).

[7] Additional useful information on after-school programs is available through the website of the Food Research and Action Center (FRAC) — [http://www.frac.org].

[8] There is a noticeable gap between the proportion of children whose families apply and who are certified eligible for free or reduced-price lunches and the proportion of children who actually participate. Based on estimates from the 1999-2000 school year, at least 10% more students are certified eligible than participate.

[9] Additional useful information about the School Breakfast program may be found at the website of the Food Research and Action Center (FRAC) — [http://www.frac.org] —specifically, a FRAC publication entitled *School Breakfast Scorecard 2005*.

[10] As with the School Lunch program, schools may opt to offer all meals free and not make free/reduced-price eligibility determinations.

[11] Under this program, a few *adult day care centers* — some 1,600 sponsors with about 2,600 sites serving 100,000 persons in FY2005 — also receive subsidies for meals and snacks served to elderly and chronically impaired disabled adults under the same basic terms as child care centers.

[12] At state option, subsidies for centers also may be calculated according to the family income demographics of the center — granting a standard subsidy for each meal/snack that is "weighted" (or "blended") to reflect the family income makeup of the center's children, or weighting total payments to a center by its family income makeup.

[13] Under FNS policies, any funding for-profit centers receive that includes *some* Title XX contribution meets this requirement — including funding sources that "pool" Title XX money with other funds (e.g., Child Care and Development Block Grant (CCDBG) funding).

[14] Also, see (1) CRS Report RL31578, *Child Nutrition and WIC Legislation in the 106th and 107th Congresses* and (2) CRS Report RL33299, *Child Nutrition and WIC Legislation in the 108th and 109th Congresses*, both by Joe Richardson.

[15] More useful information may be found at the website of the Food Research and Action Center (FRAC) — [http://www.frac.org].

[16] However, like the child care component of the program, the day care home component of the CACFP does not regulate fees charged parents. Provision of free or reduced-price meals/snacks is not required, although day care homes may adjust their fees to take federal subsidies into account.

[17] In all cases, children of a day care home *operator* must have family income below 185% of the federal poverty income guidelines (the income cutoff for reduced-price school meals) in order to have their meal/snack subsidized.

[18] Additional useful information about the Summer Food Service program may be found at the website of the Food Research and Action Center (FRAC) — [http://www.frac.org] —notably, a FRAC publication entitled: *Hunger Doesn't Take a Vacation: A Status Report on the Summer Food Service Program for Children (2005, Fifteenth Edition).*

[19] Documentation requirements are *not* applied to public and private nonprofit sponsors in 26 states and Puerto Rico under simplified summer program (so-called "Lugar") rules. Sponsors in these areas can receive the maximum federal subsidies without documenting all their costs.

[20] Also, see the discussion of fresh fruit and vegetables in the school lunch section of this report. For important supplemental information, see CRS Report RS20235, *Farm and Food Support under USDA's Section 32 Program*, by Geoffrey Becker.

[21] One state (Kansas) receives cash in place of commodity assistance. In a limited number of cases, schools, in lieu of commodities, receive cash payments or "commodity letters of credit" (CLOC) to purchase commodities themselves.

[22] The *Commodity Supplemental Food program (CSFP)* — the predecessor to the WIC program — provides federally purchased commodity food packages to low-income elderly, as well as women, infants, and children. It operates through more than 145 local projects in 32 states, the District of Columbia, and two tribal areas. While the overwhelming majority of its recipients (90%) are now elderly persons, some 50,000 women, infants, and children were assisted in FY2005.

[23] Very useful additional information on the WIC program is available from a recent report by the Agriculture Department's Economic Research Service — *The WIC Program: Background, Trends, and Issues*.

[24] The separate Defense Department WIC program is funded with Defense Department money.

[25] Two states (Mississippi and Vermont) provide WIC food items directly to recipients (not using vouchers), and several Indian tribal agencies directly distribute some items such as infant formula. Five state agencies — Nevada, New Mexico, Ohio, Texas, and Wyoming (statewide) — have implemented electronic benefit transfer (EBT) systems that provide WIC benefits through EBT debit cards rather than paper vouchers.

[26] Approximately two-thirds of NSA expenses are for nutrition-related service activities like nutrition education, breastfeeding support, and nutrition risk evaluations. The remainder (roughly 10% of total program costs) represents traditional administrative costs (e.g., income eligibility determinations, issuing and redeeming vouchers).

[27] Recently, new state rebate/discount contracts for infant formula (as previous ones expired) have yielded noticeably smaller discounts — on the order of 76%-94% (the median being 89%). For additional information on infant formula rebates, see the March 2006 Government Accountability Office (GAO) Report GAO-06-380 entitled *Food Assistance: FNS Could Take Additional Steps to Contain WIC Infant Formula Costs*.

[28] States also pursue other initiatives to control food costs, including use of "least-cost" brands and "economic" package sizes.

[29] Also see the discussion of fresh fruit and vegetables in the section of this report devoted to the School Lunch program.

[30] Money for research, studies, and evaluations is *not* included in the funding tables at the end of this report.

[31] The largest share of these new costs occurs in FY2006. For a description of the new initiatives, see CRS Report RL33299, *Child Nutrition and WIC Legislation in the 108th and 109th Congresses*, by Joe Richardson.

[32] Note: Beginning with FY2005, WIC farmers' market appropriations have been made under a separate FNS budget account — *Commodity Assistance Programs.*

In: Food Safety and Nutrition
Editors: J. Miller, P. Bonete, pp. 29-48
ISBN: 978-1-60456-335-1

Chapter 2

FOOD SAFETY: FEDERAL AND STATE RESPONSE TO THE SPINACH *E. COLI* OUTBREAK*

***Donna V. Porter*[1] *and Sarah A. Lister*[2]**
[1] Specialist in Life Sciences Domestic Social Policy Division
[2] Specialist in Public Health and Epidemiology
Domestic Social Policy Division

ABSTRACT

In September 2006, the Food and Drug Administration (FDA) and the Centers for Disease Control and Prevention (CDC) began receiving reports on clusters of patients in various states confirmed to have *E. coli* infections. By early October, 199 people in 26 states had become ill — 102 had been hospitalized, 31 had developed hemolytic uremic syndrome (HUS), a type of kidney failure, and three had died.

Escherichia coli O157:H7 (*E. coli*) is a bacterium found in animal feces that causes diarrhea and abdominal cramps within days of exposure. An infection can lead to HUS and, in some cases, death. Public health laboratories perform a type of DNA fingerprinting to determine whether a sample taken from a patient matches those taken from other patients and contaminated food during an outbreak. The time from exposure to confirmation of an *E. coli* infection can take two to three weeks.

---

* Excerpted from CRS Report RL33722, dated November 13, 2006.

As the number of infections increased, an investigation that included FDA, CDC, and state and local public health officials was launched. Starting on September 14 and continuing into early October, FDA and CDC released nearly daily statements on the status of the investigation, alerting the public to the number of cases, states with confirmed cases, spinach product recalls, agency actions, and consumer advice on consumption of spinach products. Investigators were able to trace the outbreak back to several farm fields in the Salinas Valley of California. While the investigation continues, there is evidence that nearby livestock, feral pigs or other environmental sources may have contaminated one or more of the fields.

Since the outbreak, FDA has advised growers of fresh produce that they need to develop and implement voluntary guidelines to prevent outbreaks of food-borne diseases. FDA has also announced that it will convene a public meeting on the issue once the investigation is complete. Several growers groups have called for their industry to use the best agricultural and processing practices to prevent such outbreaks, not least because losses to the industry from the spinach outbreak have been estimated at $100 million.

In October 2006, the House Committee on Energy and Commerce asked FDA to explain its role in detecting contaminated food, particularly in the recent tainted spinach case. The request sought details about the agency's food safety procedures used in emergency situations.

Both proponents and opponents of the National Uniformity for Food Act (S. 3128, H.R. 4167) have weighed in on how it would have affected the *E. coli* outbreak. Opponents believe that states' ability to act would have been compromised, while proponents claim that the legislation would not have affected state adulteration or inspection authorities.

Legislators may address the recent *E. coli* outbreak during reauthorization of the farm bill in the 110th Congress, when the proposal for a single food safety agency with increased powers may be considered.

INTRODUCTION

In September 2006, government officials were alerted to an outbreak of *E. coli* O157:H7 infections associated with the consumption of tainted fresh spinach. For several weeks, the Food and Drug Administration (FDA) and the Centers for Disease Control and Prevention (CDC) investigated the situation as additional cases were identified nearly daily. This report details the events as they unfolded, and includes the number of cases, the detection of the first case, and the process by which relevant agencies acted as subsequent cases were reported. This report will be updated in response to any further developments.

Background

The public first became aware of a new deadly strain of *E. coli* in 1982 during an outbreak associated with ground beef. *Escherichia coli* is a bacterium normally found in the intestines of humans and animals. Most strains of *E. coli* are harmless. Several strains, including serotype O157:H7, may cause serious illness in humans, though they are frequently found in livestock feces, particularly in cattle manure. In humans, infection with *E. coli* O157:H7 can cause diarrhea that is often bloody and accompanied by abdominal cramps. Fever may occur. Symptoms usually develop in two to four days, but may emerge as quickly as a day or up to a week after exposure. Healthy adults can generally recover completely from infection within a week. Some individuals, however, especially young children and the elderly, can develop hemolytic uremic syndrome (HUS) as a result of infection. HUS can lead to serious kidney damage and even death. In terms of treatment, antibiotics are not indicated and may be harmful. Due to the severity of illness, treatment of these infections often requires hospitalization. Patients who experience kidney failure may need dialysis. For the remainder of this report, "*E. coli*" will refer to the *E. coli* O157:H7 strain implicated in the spinach outbreak.

The Process of *E. coli* Case Confirmation

The cases of *E. coli* linked to spinach consumption, as discussed in this report, are "confirmed," meaning that victims have been shown, by laboratory analysis of specimens, to be infected with the organism. There are a number of reasons why confirmation may not occur for all victims of an outbreak, though for serious illnesses such as *E. coli*, the proportion of cases that are investigated with laboratory testing is generally higher than with milder foodborne illness. The time from the beginning of a patient's illness to confirmation of whether the patient is part of an outbreak typically takes from two to three weeks. In the case of the *E. coli* outbreak in spinach, the average time for confirmation of cases was about 15 days.[1]

In most cases of *E. coli* infection, public health laboratories in states and some cities perform a type of DNA fingerprinting on *E. coli* samples. Investigators determine whether the DNA fingerprinting pattern of the bacterium from one patient is the same as that from other infected patients and from contaminated food. Bacteria with the same DNA fingerprint are likely to have come from the same source.

A series of steps takes place between the point when a patient is infected and the point when public health officials can confirm whether the patient is part of an *E. coli* outbreak. As a result, there is generally a two- to three-week delay between the start of the illness and confirmation of the patient's connection to the outbreak.

This series breaks down as follows: In the case of *E. coli*, an incubation period from the time of eating contaminated food to the beginning of the first symptoms is typically two to four days. The time from the first symptom until the person seeks medical care, when a diarrhea sample is collected for testing, is generally one to five days. The process of laboratory diagnosis, which begins when a patient provides a sample and *E. coli* is subsequently obtained from the sample, usually takes one to three days. The time required to ship *E. coli* bacteria from a laboratory to state public health authorities who will perform DNA fingerprinting may take up to a week, depending on the transportation system within a state and the distance between a clinical laboratory and public health department. The time required for state public health officials to perform DNA fingerprinting on an *E. coli* sample and compare it with an outbreak pattern is ideally one day. However, with limited staff and space in public health laboratories during a period when other emergencies may occur, the process can take up to four days.

State health departments typically report laboratory-confirmed cases of foodborne illness to CDC on a regular basis. During serious interstate outbreaks such as the *E. coli* outbreak linked to spinach, state health departments would typically notify CDC of newly confirmed cases on a daily basis, to facilitate a swift nationwide investigation. In general, disease reporting by states to CDC is voluntary, but it may be required as a condition of federal funding for certain state public health systems, such as the PulseNet system described below.

The Systems Used to Monitor Foodborne Illnesses

The tracking and reporting of foodborne illnesses are conducted in several ways:

- PulseNet is a national network of public health and food laboratories coordinated by the CDC. The network consists of labs in state and local health departments and federal agencies (CDC, FDA, and the U.S. Department of Agriculture, or USDA). PulseNet experts perform standardized DNA fingerprinting of foodborne disease-causing bacteria by *pulsed-field gel electrophoresis* (PFGE). PFGE can be used to distinguish among strains of organisms — for example, *E. coli* O157:H7, *Salmonella*, *Shigella*, *Listeria* and *Campylobacter*.[2]

- The OutbreakNet is a group of state public health officers who investigate foodborne disease outbreaks and share information throughout an outbreak.
- CDC's Health Alert Network (HAN) is a national program that provides vital health information and the infrastructure to support dissemination of essential information to public health and medical professionals at the state and local levels.
- The Epidemic Information Exchange (Epi-X) is CDC's secure, Web-based communications network that serves as a communications exchange between CDC, state and local health departments, poison control centers, and other public health professionals.

All of these systems have been used in the investigation of the recent *E. coli* outbreak.

TIMELINE OF CASES, RECALLS, AND AGENCY ACTIONS[3]

Federal and State Tracking and Investigation

September 8, 2006

According to FDA and CDC reports, the agencies were first alerted on September 8 to four cases of hemolytic uremic syndrome (HUS) in a call from Wisconsin's state epidemiologist. CDC began an investigation, working collaboratively with state health departments and FDA, to detect infections, identify the cause of the infections, and provide information to the public and health care providers on the treatment and prevention of *E. coli* O157:H7 infections.

September 12, 2006

On September 12, the PulseNet system confirmed that the *E. coli* O157:H7 strains from the patients in Wisconsin all had the same DNA fingerprint pattern, and also identified the same pattern in some *E. coli* patients from other states.

September 13, 2006

By September 13, CDC officials were alerted by epidemiologists in Wisconsin and Oregon that fresh spinach was the suspected source of small clusters of *E. coli* cases. The same day, Wisconsin and Oregon epidemiologists were contacted by New Mexico epidemiologists about a cluster of *E. coli* infections that had also been associated with fresh spinach consumption. At this point, the association of illness with spinach consumption was based only on epidemiologic evidence: the common finding, among victims who were interviewed, of a history of recent spinach consumption.

September 14, 2006

FDA and CDC issued the first of nearly three weeks of daily consumer alerts about an *E. coli* outbreak in several states that was believed to be associated with the consumption of fresh produce. At the time, preliminary epidemiological evidence pointed to bagged fresh spinach as the possible cause of the outbreak. FDA advised consumers to avoid consuming bagged spinach. (It is notable that this product-wide advisory was made based solely on epidemiologic evidence, since at this point, the outbreak organism had not yet been identified in any spinach products.) The agencies advised individuals who believed that they had experienced symptoms of illness after consuming bagged spinach to contact their physicians, and physicians were urged to report suspected cases of *E. coli* infection to local and state public health officials as soon as possible.

The federal agencies reported that they were working with state and local agencies to determine the cause and scope of the outbreak. Eight states had reported illnesses: Connecticut, Idaho, Indiana, Michigan, Oregon, New Mexico, Utah, and Wisconsin. A total of 50 infected individuals had been identified — among them were eight cases of HUS, one death, and multiple hospitalizations. The date range of infection was estimated to be from August 25 to September 3, 2006.

September 15, 2006

On September 15, FDA and CDC issued announcements that the outbreak of *E. coli* in multiple states had been associated with the consumption of fresh spinach and fresh spinach-containing products. The statements indicated that Natural Selection Foods LLC of San Juan Bautista, California, had launched a voluntary recall of products that contained spinach marked with "best if used by" dates of August 17 through October 1, 2006. (FDA does not have authority to mandate recalls of most of the foods it regulates, including fresh produce.) FDA reported that it was investigating whether eight other companies and their brands were involved. The products under investigation, and subject to an expanded consumer advisory, included spinach and any salad blend containing spinach intended for retail or food service (restaurant and institutional) use. CDC also noted that 94 cases of illness had been reported, and that 29 people (31%) had been hospitalized, 14 (15%) had developed HUS, and one had died.

By this point, California, Kentucky, Maine, Minnesota, Nevada, New York, Ohio, Pennsylvania, Tennessee, Virginia, Washington, and Wyoming had also reported human infections to CDC, bringing the total number of affected states to 20. Spinach from presumptively affected lots was also reported to have been distributed in Canada and Mexico. The agencies also indicated that they, or state officials, were testing available packages of spinach consumed by victims of *E. coli* infection.

September 16, 2006

On September 16, CDC reported that 102 cases of illness due to *E. coli* infection had been confirmed, and that 52 people (51%) had been hospitalized, 16 (16%) had developed HUS, and one had died. Tennessee was removed from the list of states that had confirmed cases because a case originally attributed to the state had actually occurred in Kentucky. FDA advised consumers to avoid fresh bagged spinach, and issued company recall information.

FDA also announced that it was expanding its lettuce safety initiative to cover spinach. In response to repeated *E. coli* outbreaks associated with fresh lettuce, the agency had advised growers in November 2005 of its concerns about the safety of fresh greens, and the need for continued efforts to assure good agricultural and processing practices within the industry.[4] The California Department of Health Services had expressed similar concerns in a letter to California growers in January 2006.[5]

September 17, 2006

On September 17, CDC reported that 109 cases of *E. coli* infection had been confirmed, and that 55 people (50%) had been hospitalized, 16 (15%) had developed HUS, and one had died. FDA announced that a second voluntary recall was under way by the company River Ranch of Salinas, California, which was voluntarily recalling packages of spring mix obtained in bulk from Natural Selection Foods. The FDA's report listed all River Ranch and Natural Selection brands.

September 18, 2006

On September 18, CDC reported that 114 cases of *E. coli* had been confirmed, and that 60 people (53%) had been hospitalized, 18 (16%) had developed HUS, and one had died. CDC added Illinois and Nebraska to the list of states with confirmed cases, bringing the total to 21. The rest of the information reported by CDC and FDA, which was repeated in subsequent notices, was the same as on previous days: consumer advice, symptoms of illness, the two recalls, the lettuce safety initiative and the ongoing investigation.

September 19, 2006

On September 19, CDC reported that the number of *E. coli* cases reported had risen to 131, and that 66 people (50%) had been hospitalized, 20 (15%) had developed HUS, and one had died. FDA stated that products containing tainted spinach had been distributed to Taiwan, as well as to Canada and Mexico (as noted above), but that no illnesses had been reported by those countries.

September 20, 2006

On September 20, CDC reported that the number of reported cases had increased to 146, and that 76 people (52%) had been hospitalized, 23 (16%) had developed HUS, and one had died. Arizona and Colorado had been added to the list of states with confirmed cases, bringing the total to 23. FDA expanded its consumer alert to include fresh spinach in bagged products, spinach in a clamshell, and spinach from farmers' markets. The agency also indicated that it had found no evidence that spinach that was frozen, canned, or an ingredient in pre-made meals manufactured by food companies was tainted. FDA reported a third recall that had been announced by RLB Food Distributors, L.P., of West Caldwell, New Jersey, for certain listed salad products that may have contained spinach with an "enjoy thru" date of September 20, 2006.

The agencies' alerts also noted that the New Mexico Department of Health had announced that through DNA fingerprinting, it had matched a strain of *E. coli* from a victim's package of spinach with the outbreak strain taken from infected patients. This marked the first laboratory confirmation of the link between spinach consumption and *E. coli* infections.

September 21, 2006

On September 21, CDC announced that 157 cases of *E. coli* illness had been reported, and that 83 people (52%) had been hospitalized, 27 (17%) had developed HUS, and one had died. FDA reported that it was working closely with the state of California, since it had been determined that the contaminated spinach had come from fields in the California counties of Monterey, San Benito, and Santa Clara. FDA alerts reassured consumers that processed spinach (frozen and canned) and other produce from those counties were not implicated in the outbreak. The same day, CDC held a Clinician's Outreach and Communication Activity (COCA) conference call with 800 people, who heard experts provide an overview of the current outbreak and FDA's investigation, the nature of *E. coli* as a pathogen, and treatment options for patients exhibiting symptoms of the infection.

September 22, 2006

On September 22, CDC indicated that 166 cases of illness had been reported, and that 88 people (53%) had been hospitalized, 27 (16%) had developed HUS, and one had died. Maryland and Tennessee were added to the list of states with confirmed cases, bringing the total to 25. FDA reported working closely with the state of California in the three counties in which the tainted spinach may have been grown — investigators were attempting to narrow the geographic area suspected of being the source of the outbreak. FDA repeated the consumer notice that processed spinach, spinach grown elsewhere in the United States, and other produce grown in the three implicated counties were safe to eat, and that the food industry was working to get fresh spinach back on the market.

September 23, 2006

On September 23, CDC announced that 171 cases of *E. coli* infection had been reported, and that 92 people (54%) had been hospitalized, 27 (16%) had developed HUS, and one had died. In addition, FDA reported two more voluntary recalls for products containing spinach supplied from Natural Selection Foods of California. Triple B Corporation, doing business as S.T. Produce of Seattle, Washington, recalled its fresh spinach products with "use by" dates of August 22 through September 20, 2006, that had been distributed to retail stores in Idaho, Montana, Oregon, and Washington, and sold in hard plastic clamshell containers. Pacific Coast Fruits Company of Portland, Oregon, recalled products that might contain spinach with "use by" dates of September 20, 2006, or earlier, and on pizza products with dates of September 23, 2006, or earlier. The company reportedly had stopped making all products with spinach supplied from California on September 14, 2006. Its products were shipped to Alaska, Idaho, Oregon, and Washington. For the 25 states with confirmed cases, the number of cases was reported for each state. Hong Kong was added to the list of places outside the U.S. that had received the affected products.

September 24, 2006

On September 24, CDC stated that 173 cases had been reported, and that 92 people (53%) had been hospitalized, 28 (16%) had developed HUS, and one had died. The agencies indicated that the Utah Department of Health and the Salt Lake Valley Health Department had confirmed that *E. coli* O157:H7, the same strain as the one associated with the outbreak, had been found in a bag of Dole baby spinach purchased in Utah with a "use by" date of August 30, 2006. The tests were conducted by the Utah Public Health Laboratory.

September 25, 2006

On September 25, CDC stated that 175 cases had been reported, and that 93 people (53%) had been hospitalized, 28 (16%) had developed HUS, and one had died. FDA reported that tainted products had been distributed to Iceland.

September 26, 2006

On September 26, CDC reported that the number of cases had increased to 183, and that 95 people (52%) had been hospitalized, 29 (16%) had developed HUS, and one had died. West Virginia was added to the list of affected states, bringing the total to 26. CDC and FDA also announced that Canada had reported one confirmed case of *E. coli* O157:H7 that matched the outbreak strain. In addition, they noted that Pennsylvania Department of Health had reported that the outbreak strain of *E. coli* O157:H7 had been isolated from a bag of baby spinach.

September 28, 2006

On September 28, CDC reported that 187 cases infected with the outbreak strain of *E. coli* had been reported, and that 97 people (52%) had been hospitalized, 29 (16%) had developed HUS, and one had died.

September 29, 2006

On September 29, FDA announced that it had determined that spinach implicated in the *E. coli* outbreak had been traced to Natural Selection Foods LLC of San Juan Bautista, California. The determination was based on epidemiologic and laboratory evidence obtained from multiple states (Colorado, Ohio, Wisconsin, Nevada, Pennsylvania, Utah, New Mexico, and Illinois) and analyzed by CDC, and on product distribution information. One company had recalled its product on September 15 and four others had instituted secondary recalls, because they received the recalled product from Natural Selection Foods. The statement indicated that FDA, the state of California, CDC, and the US. Department of Agriculture (USDA) were continuing to investigate the cause of the outbreak through ongoing inspections and sample collections at facilities, in the environment (including irrigation water sources), and through studies of local animal management and water use.

The FDA update indicated that the Grower Shipper Association of Central California, the Produce Marketing Association, the United Fresh Produce

Association, and the Western Growers Association had agreed to develop a voluntary plan to improve the safety of fresh produce. FDA and the state of California, however, have not ruled out the possibility of instituting regulatory requirements in the future. In addition, FDA intends to convene a public meeting later in the year to address the larger issue of food-borne illnesses linked to leafy greens once the current investigation is complete.

October 3, 2006

On October 3, CDC reported 192 cases of illness, and that 98 people (51%) had been hospitalized, 30 (16%) had developed HUS, and one died.

October 5, 2006

On October 5, FDA announced that the U.S. Attorney for the Northern District of California had issued a statement on the execution of two search warrants — for Growers Express in Salinas and Natural Selection Foods in San Bautista, California — in connection with the outbreak of *E. coli* O157:H7 that FDA had traced to spinach grown in the Salinas area. The U.S. Attorney stated that there was no indication at that time that leaf spinach had been deliberately or intentionally contaminated. FDA stated that it was working with the U.S. Attorney's office and the Federal Bureau of Investigation (FBI) to determine the facts behind the outbreak, particularly allegations that certain spinach growers and distributors may have failed to take all necessary precautions to ensure that their spinach was safe before it was placed in interstate commerce. The number of reported cases of illness had not changed.

October 6, 2006

On October 6, CDC announced that the number of reported cases had increased to 199 (see figure 1, below) — 102 people (51%) had been hospitalized, and 31 (16%) had developed HUS. On this date, CDC also announced that a total of three people had died in the outbreak — the single case announced earlier, an elderly woman in Wisconsin; and two more victims with *E. coli* infections that matched the outbreak strain on DNA fingerprinting, a child in Idaho and an elderly woman in Nebraska. CDC also reported a fourth suspicious death in an elderly woman in Maryland, but samples to confirm infection by DNA fingerprinting were not available. The two confirmed deaths announced on this date had occurred in September, but required additional time to link to the outbreak by DNA fingerprinting. CDC reported additional findings as follows:

- Of the individuals affected by the outbreak, 141 (71%) were female and 22 (11%) were children under the age of five.
- The proportions of individuals in each age group who developed HUS included 29% of children under age 18, 8% of those aged 18 to 59, and 14 % of those who were 60 years or older.
- The cases were spread across 26 states: Arizona (8), California (2), Colorado (1), Connecticut (3), Idaho (7), Illinois (2), Indiana (10), Kentucky (8), Maine (3), Maryland (3), Michigan (4), Minnesota (2), Nebraska (11), Nevada (2), New Mexico (5), New York (11), Ohio (25), Oregon (6), Pennsylvania (10), Tennessee (1), Utah (19), Virginia (2), Washington (3), West Virginia (1), Wisconsin (49), and Wyoming (1). (See figure 2, below).

In addition, FDA had determined, based on recall audits, that on September 15, Kenter Canyon Farms, Inc., of Sun Valley, California, had instituted a voluntary recall of repackaged spinach as part of the nationwide recall of Natural Selection Foods. The recalled product was only distributed in California, and carried an expiration date of September 20, 2006.

October 12, 2006

On October 12, FDA and the state of California announced test results from the field investigation of the outbreak of *E. coli* O157:H7 in spinach. Samples of cattle feces taken from one of the four implicated ranches tested positive based on matching DNA fingerprints for the same strain that sickened 199 individuals. The announcement stated that while this finding was significant, it was just one aspect of the trace-back investigation that was ongoing for FDA, the state of California, CDC, and USDA. As of November 8, 2006, this alert was the agencies' most recent. (Readers should note that later news reports[6] have indicated that there were a total of 204 confirmed cases, but this number had not been verified by any available agency documents as of November 8, 2006.)

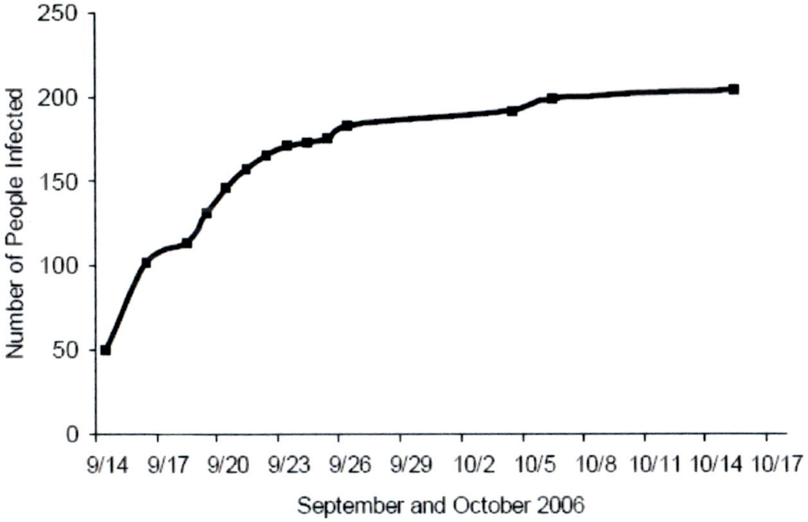

Source: CDC and FDA data; compiled by CRS.

Figure 1. Cumulative Number of *E. coli* Cases Reported by CDC.

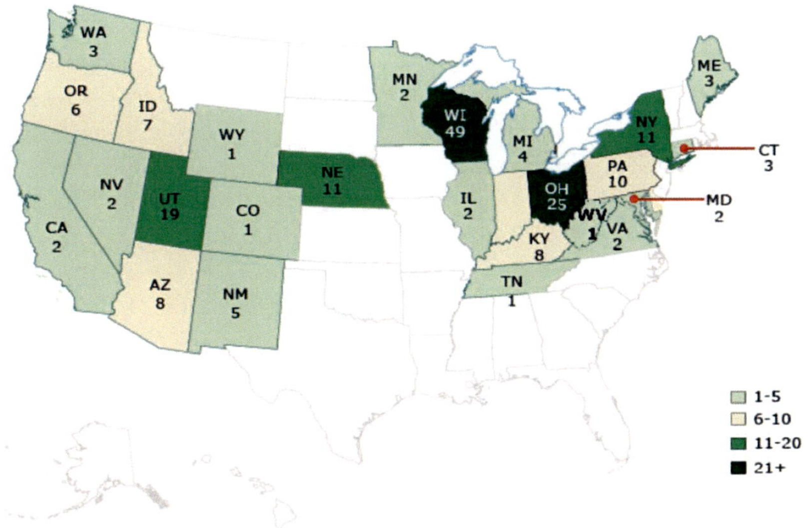

Source: Information provided by CDC. Map adapted by CRS.

Figure 2. Number of Confirmed *E. coli* Cases by State as of October 6, 2006.

Developments Following the Agency Alerts on the Outbreak

On October 22, 2006, a statement issued by the United Fresh Produce Association indicated that it understood that FDA had eliminated all concerns about spinach grown anywhere outside of the three counties in California.[7] The group indicated its commitment to determining the specific source of the contamination and working to prevent future outbreaks, including the use of the best agricultural practices in the field and the strongest possible Hazard Analysis and Critical Control Points (HACCP) programs in processing facilities.

The Center for Science in the Public Interest (CSPI) petitioned the state of California to adopt stricter food safety procedures because the state has been at the center of the *E. coli* outbreak, and has experienced other problems with contamination of leafy greens.[8] The petition calls for the California Department of Health Services to establish mandatory regulations relating to manure and water safety on farms, similar to those required of the meat and poultry industries. CSPI suggested that the use of manure as fertilizer should be prohibited during growing season, and that only drinkable water should be used in produce processing facilities.

On October 29, 2006, a news article reported that the investigation of the *E. coli* spinach outbreak had been thorough, although there were no final answers yet.[9] The article indicated that the codes printed on the bags of spinach had reportedly led detectives to discover when it had been bagged — on August 15 — and other specific details, including worker shift and packing line. Exhaustive testing of the plants' equipment and water supply over several weeks turned up none of the implicated bacteria, according to government officials and company representatives. Company records led investigators to the fields of nine farms in the three counties where the spinach packed on August 15 had been grown. Coding on additional bags of contaminated spinach allowed investigators to narrow the search to four fields. FDA reported that the strain of *E. coli* had been found in manure on a cattle ranch in the Salinas Valley, within a mile of spinach fields. Investigators combed the fields for more samples, including wildlife and cattle feces, stream water, and spinach leaves. Six samples taken from the ranch tested positive for the *E. coli* strain that was being sought, including one found in the gut of a feral pig killed on the property. There were also signs that feral pigs had broken through a wire mesh fence to reach the spinach, indicating that the field was a likely source of the outbreak and that the wild pigs were the probable carriers of the bacteria. However, other fields have not been ruled out as the source. Nevertheless, this investigation has provided the most specific information to date for how a microscopic organism commonly found in an animal's feces can cause illness and death in consumers thousands of miles away from the primary source of the contamination.

On October 30, 2006, the Western Growers Association proposed that farmers who grow leafy greens be subject to new food safety standards.[10] While a detailed plan has yet to be released, the statement indicated that soil, irrigation water, and farm equipment should be tested for bacteria, workers should be trained in food safety practices, and processing plants should be checked regularly. Violators could be fined or banned from shipping. The association indicated that the produce industry is subject to little oversight by state and federal regulators, an arrangement that is believed to have led to the *E. coli* outbreak. Losses to the spinach industry alone from the recent outbreak have been estimated to exceed $100 million.[11]

The association proposes to fund the food safety measures through the novel use of a marketing order. Marketing orders were authorized by state and federal legislation in 1937 as a way to compel producers of a particular crop to follow a common set of rules and pay fees used to fund programs. While marketing orders have never been used to enforce food safety, California Department of Food and Agriculture officials have said that such procedures appear to be within the scope of the law. No information is available on the cost of implementing such a program. Although farmers have to approve marketing orders, they do seem to support the call for increased regulation. The September 2006 outbreak represented the ninth outbreak in California in a decade, and the 20th report of an *E. coli* outbreak in lettuce or leafy greens nationwide since 1995. FDA has reportedly pressed the lettuce and spinach growing industry for more than two years to follow voluntary federal guidelines to prevent outbreaks of foodborne diseases.

On November 2, 2006, a rancher in San Benito County, California, revealed that his operation was one of four farms under investigation by government agencies in connection with the September 2006 *E. coli* outbreak.[12] The rancher indicated that his operation did not grow or process the spinach in question, but that he rented fields to two tenants, one of whom was still under investigation as of November 8. Also on November 2, the Canadian government lifted its ban on U.S. spinach that is grown anywhere outside of San Benito and Monterey counties. Mexico had reportedly lifted its ban on California-grown lettuce in October.

On November 2, 2006, several major supermarkets told growers that they had six weeks to establish new safety rules to prevent *E. coli* outbreaks.[13] The consortium of stores (Vons, Ralph's, and Albertsons grocery chains, and the Costco Wholesale Corporation) wanted growers to work with federal regulators, academia, and industry research scientists to standardize food safety requirements. The group also called for a process for updating food safety rules in response to the emergence of research findings on how diseases are spread from the farm to the dinner table. If growers failed to achieve more stringent and enforceable farming practices, the consortium indicated that it was prepared to set up its own certification system.

RELATED CONGRESSIONAL ACTIVITIES

On October 24, 2006, the House Committee on Energy and Commerce asked FDA to explain its role in detecting contaminated food, particularly the *E. coli* strain that tainted spinach in September 2006.[14] The request asked for detailed information about the agency's food contamination preparedness and deterrence assessment, conducted in 2004, and how the assessment was being used in emergencies like the 2006 *E. coli* outbreak. The committee's letter asked the agency for the types of food commodities selected for vulnerability assessments under its 2004 assessment, known as the FDA Security Surveillance Assignment (FSSA), and how the data were collected and analyzed. (FSSA was conducted to determine the safety of food at special security events, such as the G-8 Summit and the national political conventions.)

The recent *E. coli* outbreak led opponents of the National Uniformity for Food Act (S. 3128, H.R. 4167), which is scheduled to come before the Senate in the closing days of the 109th Congress, to call for a rejection of the bill. The Association of Food and Drug Officials (AFDO) has indicated that states' ability to react to future bacterial outbreaks would be compromised. While FDA does not have authority to mandate recalls for most of the foods it regulates, the states generally do have this authority. Proponents of the bill argue that the legislation would not affect state adulteration or inspection authorities. Supporters believe that state food regulators are simply wrong to say that the legislation standardizing food warning labels would have hindered states' ability to recall spinach. See CRS Report RL33559, *Food Safety: National Uniformity for Food Act*, by Donna V. Porter, for further information.

The House and Senate Agriculture Committees may address the recent outbreak of foodborne illness that occurred as a result of spinach contaminated by *E. coli,* possibly from animal manure, when they take up reauthorization of the farm bill in the 110th Congress. Some members have been longtime critics of the division of food safety responsibilities between USDA, which regulates meat, poultry, and processed egg products, and FDA, which regulates all other food products. A proposal for a single food safety agency with increased powers may become part of the debate.

While not mentioned in FDA's reports of its investigation, it is likely that the investigation was streamlined as a result of certain new authorities granted to FDA in comprehensive bioterrorism and public health preparedness legislation in the 107th Congress. P.L. 107-188, the Public Health Security and Bioterrorism Preparedness and Response Act of 2002, gave FDA the authority to require that food processing facilities register with the agency, which had not been previously required. In addition, processors were required to maintain records that FDA could use to facilitate product trace-backs during outbreak investigations, including in their records information about the source of ingredients coming into their facilities ("one step back") and the customers to whom they shipped products ("one step forward").

FINAL OBSERVATIONS

The federal and state systems for tracking and investigating the bacterial outbreak seem to have worked well once the clusters of *E. coli* infections were recognized in several states. Investigators were able to trace back to the likely location and source of the problem through a mix of epidemiologic and laboratory investigation, and multi-agency coordination. The FDA and CDC provided alerts to the public and to health officials, and continued to do so through daily updates. In addition, CDC held a conference call for 800 health care professionals on the investigation of the outbreak, *E. coli* testing, and treatment of patients.

A weak point in the system seems to be how best to intercept *E. coli* contamination before it enters the food chain. While procedures are in place to enhance the safety of meat products, the voluntary federal guidelines for leafy greens seem to be ineffective or not fully implemented by growers and processors. Plans to develop a more effective system announced by the produce industry may address this weakness in the food chain. However, mandatory requirements may need to be implemented by regulators to assure the public that leafy greens are safe for consumption. While funding through market orders has been proposed by one group, the question of whether this source of funding or another source is used remains to be resolved. Oversight of the produce safety system will likely be needed.

In the final analysis, the specific cause of the California spinach *E. coli* outbreak may never be known. Investigators were able to develop a fairly good idea of how the contamination and outbreak occurred, thanks to identifying information on bagged spinach, and DNA fingerprinting technology. While livestock and feral pigs were shown to be carriers of the implicated *E. coli* strain, the precise pathway by which the spinach became contaminated is as yet unclear. Best agricultural practices and HACCP in processing operations are likely to help in preventing such outbreaks in the future.

REFERENCES

[1] CDC, "Timeline for Reporting of Cases with the Outbreak Strain of E. coli O157," Sept. 19, 2006, at [http://www.cdc.gov/foodborne/ ecolispinach/ reportingtimeline.htm].

[2] CDC, PulseNet home page, at [http://www.cdc.gov/pulsenet/index.htm].

[3] The data in this section were taken from the daily alerts issued by FDA and CDC, available at the following sites: [http://www.fda.gov/oc/opacom/ hottopics/spinach.html] and [http://www.cdc.gov/foodborne/ecolispinach/].

[4] FDA, "Nationwide *E. coli* O157:H7 Outbreak: Questions and Answers," Sept. 16, 2006, at [http://www.cfsan.fda.gov/~dms/spinacqa.html].

[5] California Department of Health Services, "CDHS *E. coli* Information Website," at [http://www.dhs.ca.gov/opa/ecoli/].

[6] "Turning over a new leaf: Produce group wants farmers to foot the bill for inspections." *The Sacramento Bee*, October 31, 2006.

[7] "Escherichia coli O157:H7 outbreak investigation narrows," Medical Letter on the CDC and FDA via NewsRx.com and NewRx.net, Oct. 22, 2006.

[8] Center for Science in the Public Interest, "California urged to monitor farms for food safety," Oct. 25, 2006, at [http://www.cspinet.org/new/ 200610251.html].

[9] "Spinach probe most thorough ever, but no clear answers yet," *The Associated Press*, State and Local Wire, Oct. 29, 2006.

[10] See: Western Growers Association, "Western Growers Board Takes Action To Require Mandatory Food Safety Practices," press release, Oct. 30, 2006, at [http://www.wga.com/ public/active/siteBuilder/templateNews Release Popup.php?id=70]; and "Turning over a new leaf: Produce group wants farmers to foot the bill for inspections." *The Sacramento Bee*, Oct. 31, 2006.

[11] Ibid.

[12] One of four farms under investigation named. *The Californian* (Salinas, California), Nov. 2, 2006.

[13] "Grocers enter produce-safety debate: Big supermarket chains tell growers they have six weeks to create rules to avoid E.coli outbreaks." *The Los Angeles Times*, November 2, 2006.

[14] House Committee on Energy and Commerce, "Oversight letter to the FDA concerning the safety of the U.S. food supply and the adequacy of the FDA's food safety and food security efforts,"letter to Dr. Andrew C von Eschenbach, FDA Commissioner, Oct. 24, 2006, at [http://energycommerce.house.gov/108/letters/letters.htm].

In: Food Safety and Nutrition ISBN: 978-1-60456-335-1
Editors: J. Miller, P. Bonete, pp. 49-70

Chapter 3

CHILD NUTRITION AND WIC LEGISLATION IN THE 108TH AND 109TH CONGRESSES*

Joe Richardson
Domestic Social Policy Division

ABSTRACT

Child nutrition programs (e.g., school meal programs, summer food service, child care food programs) and the Special Supplemental Nutrition Program for Women, Infants, and Children (the WIC Program) are subject to periodic comprehensive reviews, when appropriations and other authorities expire and have to be reauthorized. They were up for reauthorization review in the 108th Congress, and the only substantial child nutrition-WIC legislation in the 108th Congress, and so far in the 109th Congress, has been the 2004 reauthorization law — the Child Nutrition and WIC Reauthorization Act of 2004, P.L. 108-265, enacted June 30, 2004.

The 2004 law extended virtually all expiring authorities through FY2009 and contained important, but incremental, changes in child nutrition programs and the WIC program; the Congressional Budget Office estimated that it will generate net new spending totaling about $230 million through FY2009. Its major feature was a set of amendments aimed at improving the integrity and administration of the school meal programs. Significant changes were made in procedures relating to the way children's eligibility for free and reduced-price school meals is certified and verified, and new initiatives to upgrade schools' administration of their meal programs were

* Excerpted from CRS Report RL33299, dated May 23, 2006.

put in place. However, minimal revisions were made to the school meal programs themselves — for example, expansion of eligibility for homeless, runaway, and migrant children, loosened rules for certain higher school breakfast subsidies — and a major proposal to phase in higher income eligibility limits for free school meals was limited to an *authorization* for a pilot project. Relatively minor amendments also affected the Summer Food Service and Child and Adult Care Food programs — for example, making permanent and expanding coverage of "Lugar" rules facilitating participation by summer program sponsors and making permanent and nationally applicable a rule loosening Child and Adult Care Food program eligibility rules for for-profit child care centers.

Another area of concern addressed by the reauthorization law was nutrition, health, and nutrition education. Here, the biggest initiative was a requirement that all schools participating in school meal programs establish locally designed "wellness policies" to set nutrition, physical activity, and other goals and strategies for meeting them. Coupled with it were (1) *authorizations* for new nutrition education efforts, (2) an expansion of the program offering free fresh fruit and vegetables in selected schools, and (3) significant changes in food safety rules.

Finally, a large number of revisions were made to the law governing the WIC program. The most important among them were amendments aimed at strengthening rules that help contain food costs incurred by the program; these included provisions placing substantial limits on vendors receiving the majority of their revenue from WIC vouchers (so-called "WIC-only" stores).

BACKGROUND

Child nutrition programs and the Special Supplemental Nutrition Program for Women, Infants, and Children (the WIC program) are governed by three basic federal laws:[1]

- *The Richard B. Russell National School Lunch Act.* This law was originally enacted in 1946 as the National School Lunch Act; it was renamed in 1999. It provides authority for the School Lunch program, the Child and Adult Care Food program, the Summer Food Service program, after-school programs (those run under both the aegis of the School Lunch program and the Child and Adult Care Food program), food commodity support for child nutrition programs, training and technical assistance (including the Food Service Management Institute), compliance, program integrity, and accountability activities, and an information clearinghouse. It also includes a number of general-purpose provisions governing the operation of multiple child nutrition programs (e.g., income eligibility and inflation indexing rules, nutrition requirements for meals, general rules participating schools must follow).

- *The Child Nutrition Act.* This law was originally enacted in 1966. It provides authority for the School Breakfast program, the WIC program, the Special Milk program, assistance for state administrative expenses related to child nutrition programs, and nutrition education activities.
- *Section 32* of the act of August 24, 1935 (7 U.S.C. 612c). Section 32 authority provides funding for cash child nutrition subsidies (permanent appropriations under Section 32 are transferred to the child nutrition account annually as part of the appropriations process). It also includes additional authority (to that in the Richard B. Russell National School Lunch Act) and funding for the acquisition of food commodities for distribution to child nutrition programs (Section 32 money is used to buy and distribute surplus commodities).[2]

Prior to the 108th Congress, the most recent *comprehensive* amendments to child nutrition and WIC law were made by the 1998 William F. Goodling Child Nutrition Reauthorization Act (P.L. 105-336).[3] And, under the reauthorization schedule, the next wide-ranging review of child nutrition and WIC programs was set for 2003. However, a number of significant laws affecting the child nutrition and WIC programs were enacted in the 106th and 107th Congresses (1999 through 2002). These are covered in the immediate predecessor to this report — CRS Report RL31578, *Child Nutrition and WIC Legislation in the 106th and 107th Congresses,* by Joe Richardson. Information on the programs covered by child nutrition and WIC legislation is contained in CRS Report RL33307, *Child Nutrition and WIC Programs: Background and Recent Funding,* by Joe Richardson.

Only one significant law governing child nutrition programs (e.g., school meal programs) and the WIC program has been (or is likely to be) enacted in the 108th and 109th Congresses — the Child Nutrition and WIC Reauthorization Act of 2004 (P.L. 108-265; signed on June 30, 2004).

The Child Nutrition and WIC Reauthorization Act of 2004: P.L. 108-265

History

The authorization of appropriations for several child nutrition programs and the WIC program, as well as authority to carry out certain child nutrition activities, were scheduled to expire September 30, 2003, under the terms of the previous reauthorization law (P.L. 105-336).[4] Anticipating renewal of these authorizations, Congress began consideration of child nutrition/WIC reauthorization legislation in the Spring of 2003. However, primarily because of budget-related concerns as to how much, if any, new funding might be available to be spent on new or expanded initiatives, no action was taken in 2003 (other than hearings and the introduction of bills proposing various individual program changes). Instead, a short-term extension of expiring authorizations (P.L.108-134) was enacted, pushing most expiration dates to March 31, 2004, and giving Congress time to act.

On March 24, 2004, the House passed an amended version of its child nutrition-WIC reauthorization bill (the Child Nutrition Improvement and Integrity Act; H.R. 3873; H.Rept. 108-445), which included provisions that would have resulted in estimated net new mandatory spending of $235 million through FY2009. However, with the March 31st deadline approaching (and no definitive Senate action), another stopgap law extending most expiring authorities through June 30, 2004, was enacted — P.L. 108-211.

While there was no formal action in the Senate, the House and Senate committees of jurisdiction (the House Education and the Workforce Committee and the Senate Agriculture, Nutrition, and Forestry Committee), with the Administration's participation, informally discussed a potential final bill through March, April, May, and into June 2004. On May 19, 2004, the Senate Agriculture, Nutrition, and Forestry Committee ordered reported the Child Nutrition and WIC Reauthorization Act of 2004 (S. 2507; S.Rept. 108-279; reported June 7, 2004). This bill, as amended with unanimous consent on the Senate floor, passed the Senate on June 23, 2004. It reflected agreements that had been reached in the House-Senate discussions that had been taking place over the past several months and included many House (and Administration) proposals and perspectives. On June 24, 2004, the House passed it without objection. This final measure encompassed changes in law estimated to result in $232 million in net new mandatory spending (outlays, slightly more in budget authority) through FY2009 (the end of the reauthorization period). It was enacted as the Child Nutrition and WIC Reauthorization Act of 2004 (P.L. 108-165) on June 30, 2004.

The New Law

Much of the debate over and content of the 2004 reauthorization law was driven by the availability (or lack of availability) of funding for new initiatives or program expansions. Budget reconciliation agreements offered little new mandatory money (a total of about $230 million through FY2009), and, if House and Senate legislators wanted to fund new initiatives or expansions with mandatory dollars beyond these budget constraints, they effectively were forced to use savings from changes they made *within* child nutrition programs.[5] In the alternative, they could (and did) *authorize* appropriations for new initiatives/expansions or establish *pilot projects* rather than create or expand entitlement/mandatory programs. The final product encompassed some $395 million in new *gross* spending through FY2009; however, this was financed with an estimated $163 million in savings achieved primarily by revisions to the system for certifying and verifying eligibility for free and reduced-price school meals (see immediately below).

Perhaps the biggest issue confronted by Congress was concern over "overcertification" — the degree to which ineligible children are judged eligible for free or reduced-price school meals. Though estimates and perceptions differed as to the extent and seriousness of the problem, overcertification was widely judged a significant issue and provoked accompanying questions as to schools' administrative strength and how to ensure that *eligible* children were not disadvantaged by any legislative solution. As a result, the most important amendments in the 2004 law dealt with program integrity — improvements to rules governing certification and verification of eligibility, new procedures to upgrade administration of school meal programs, and new technical assistance and training.

As to school meal program operations, few changes were made (e.g., expansion of eligibility for homeless, runaway, and migrant children and youth, loosened rules for higher "severe need" school breakfast subsidies). And, a major (and costly) proposal to phase in higher income eligibility limits for free school meals was limited to *authorization* for a small pilot project.

Another major set of issues involved nutrition, health, and nutrition education aspects of child nutrition programs, particularly childhood obesity trends and the role of child nutrition programs and school environments (like physical activities) in dealing with this, the rules governing the offering of milk and milk substitutes, foods sold in competition with school meals, food safety, and the desire of some to substantially increase funding commitments for nutrition education. While a number of federally led options were offered, in the end, the most significant change was a requirement that all schools participating in school meal programs establish locally designed "wellness policies" to tackle obesity- and nutrition/health-related issues (like competitive foods and students' physical activity), with federal technical assistance and other support. Nutrition education provisions encompassed a series of *authorizations*, but no new funding. In related amendments, the program offering free fresh fruit and vegetables operating in selected schools in a few states was expanded and made permanent, and important changes were made to food safety rules.

Small changes also were made in the Summer Food Service and Child and Adult Care Food programs — for example, making permanent and expanding the coverage of pre-existing "simplified summer program" (so-called "Lugar") rules that make it easier for summer program sponsors to participate, making permanent and nationally applicable a pre-existing rule loosening eligibility requirements for for-profit child care centers wishing to participate in the Child and Adult Care Food program.

Finally, a large *number* of revisions were made to the WIC program, but the most significant were several aimed at strengthening rules containing food costs incurred by the program. Of these, the most controversial was a new provision placing substantial limits on vendors receiving the majority of their revenue from WIC vouchers (so-called "WIC-only" stores).

The 2004 reauthorization law extended virtually all expiring child nutrition and WIC appropriations authorizations and other authorities through FY2009.[6] In addition, it made a wide range of changes to child nutrition and WIC law as described in the following sections.

Program Integrity in School Meal Programs: Certification and Verification of Eligibility

- *Role of Local Educational Agencies (LEAs).* Stipulates that LEAs are effectively responsible for certification and verification of eligibility for free and reduced-price school meals — as opposed to local "school food authorities," which conduct school meal program day-to-day operations and also were in charge of certification and verification under pre-existing law.
- *Descriptive Materials and Communications with Families.* Requires that descriptive materials provided to parents contain notification that (1)

participants in certain other income-tested programs may be eligible for free or reduced-price school meals and (2) documentation may be requested for verification of eligibility. Also directs that any communications with families for eligibility determination or verification purposes be in an understandable format in a language that parents can understand (to the extent possible). Grants explicit permission for Internet applications and descriptive materials. Individual applications for each child in a family are, in most cases, barred.

- *Direct Certification of Eligibility.* In order to reduce the number of paper applications that need to be reviewed, by relying on verification already conducted in other venues, *requires* that LEAs use direct certification — certification of eligibility for free and reduced-price school meals (without application) by direct communication with appropriate state or local agencies — for any child who is member of a *food stamp* household. This direct certification mandate is phased in, beginning with the largest LEAs, through the 2007-2008 school year. In addition, *permissive* authority is granted to employ direct certification for children in Temporary Assistance for Needy Families (TANF) households, homeless children, children served by programs under the Runaway and Homeless Youth Act, and migrant children. [*Note:* Under pre-existing law, direct certification was *permitted* for children in food stamp and TANF households.]
- *Verification of a Sample of Applications.* For those not directly certified eligible for free or reduced-price school meals, mandates that a sample of approved applications be selected for verification of their eligibility by directly communicating with the family (or direct verification, see below). [*Note:* Prior to verifying with the family, the application must be reviewed by an individual other than the person who approved it.] The required sample size for each LEA varies, but, in most cases, the new law calls for it to be significantly larger than under pre-existing (regulatory) rules.
- *Direct Verification.* In conducting eligibility verification of the sample of approved applications, *permits* LEAs to first use direct verification — obtain and use income and program participation information about the applicant family from certain public agencies — before contacting the family. Direct verification is to be evaluated and, if found to be effective in obtaining adequate information and reducing the number of applications that must be verified by contacting families directly, may be made mandatory.

- *Individual Family Verification.* For families not directly certified or directly verified, requires LEAs to conduct verification through contact with the family. This includes a written request for information and phone contact if necessary. In the case of families not responding to a verification request, LEAs are required to make at least one additional attempt to obtain information and may use a third party to assist them in these "follow-up" activities. [*Note:* Nonresponse results in disqualification.]
- *Funding.* Provides mandatory funding (1) to assist in carrying out direct certification and verification activities and (2) to evaluate direct verification.

Program Integrity in School Meal Programs: Administrative Error Reduction

- *Federal Administrative Support.* Provides mandatory funding to the Agriculture Department to (1) provide training and technical assistance and materials related to improving program integrity and administrative accuracy in school meal programs and (2) assist state agencies in reviewing the administrative practices of local agencies. Requires the development and distribution of materials representative of "best management and administrative practices."
- *Additional Administrative Reviews.* Mandates that state agencies conduct special administrative reviews — beyond those carried out as part of the regular "Coordinated Review Effort" (CRE) compliance and accountability reviews — of local agencies that have demonstrated a high level of, or a high risk for, administrative error in application, certification, verification, meal counting, and meal claiming procedures. Those local agencies failing to meet performance criteria would be required to develop and carry out corrective action plans (with the state agency's assistance). If they fail to meet the criteria in both initial and follow-up reviews, local agencies could be subject to financial penalties (retention of funds otherwise due them).
- *Training.* Requires state agencies to provide training in administrative practices to local agency administrative personnel. The Agriculture Department may support this training or provide it directly.
- *Funds for Administrative Reviews and Training.* Provides mandatory funding to the Agriculture Department to assist state agencies in carrying

out additional administrative reviews and training in administrative practices.

- *Technology.* Requires that state child nutrition plans submitted for state administrative expense funding include information as to how technology and information management systems will be used to improve program integrity. Also requires a study of the use of computer technology to reduce eligibility certification errors and other "waste, fraud, and abuse." *Authorizes* the Agriculture Department to provide competitive grants to state agencies to award to local agencies to defray the cost of purchasing/upgrading technology and information systems.
- *Compliance and Accountability.* Increases the appropriations authorization for compliance and accountability reviews of local school meal program agencies.[7]

School Meal Programs (School Lunches and Breakfasts)[8]

- *Runaway, Homeless, and Migrant Youth.* Establishes in law (as opposed to administrative guidance) the automatic (categorical) eligibility of homeless children for free school meals. Also makes youth served by grant programs under the Runaway and Homeless Youth Act and migrant children automatically eligible for free school meals.
- *Duration of Eligibility.* Mandates that eligibility for free or reduced-price school meals remains valid for one year for most students.
- *Military Housing Allowances.* Makes permanent a rule disregarding military housing allowances for those living in "privatized" housing when judging eligibility for free or reduced-price school meals.
- *"Severe Need" School Breakfast Program Subsidies.* Removes a requirement that schools document their costs in order to receive higher "severe need" school breakfast subsidies, allowing them to receive the maximum subsidy rather than the lesser of documented costs or the severe need rate.[9]
- *Pilot Expansion of Eligibility for Free School Meals. Authorizes* a pilot project in all or parts of 5 states under which the income limit for free school meals would be raised from 130% of the federal poverty income guidelines to 185% of the guidelines (the current limit for reduced-price meals).
- *Evaluation. Authorizes* funds for annual national performance assessments of child nutrition meal service programs.

- *"Best Practices." Authorizes* funding for a review of "best practices" in the School Breakfast program so as to assist schools in addressing impediments to the growth of the program.

Nutrition, Health, and Nutrition Education

- *Local Wellness Policies.* Requires local educational agencies participating in school meal programs to establish local "school wellness" policies, as they determine appropriate. These policies would include goals for nutrition education, physical activity, and other school-based activities, nutrition guidelines for *all* foods available on campus during the school day, a plan for measuring implementation of the policy, and involvement of parents, students, school and school meal administrators, and the public. The Agriculture Department is directed to make available information and technical assistance to support this initiative and is given mandatory funding to provide it.
- *Nutrition Education/Promotion.* Replaces (unfunded) authority for a nutrition education and training program with *authorization* for grants to states to implement "team nutrition networks" that support nutrition education through a wide variety of means, such as the use of nutrition education messages, promotion of active lifestyles, pilot projects, model curricula, data gathering, health and physical activity guidelines. *Authorizes* funds to promote improved nutrition through the dissemination and use of nutrition messages and materials developed by the Agriculture Department. *Authorizes* grants to entities with expertise in health education programs for limited-English-proficient individuals to enhance obesity prevention activities. Requires the Agriculture Department to issue recommendations for the increased consumption of foods in school meal programs based on the most recent Dietary Guidelines for Americans. *Authorizes* grants to local educational agencies to create healthy school nutrition environments, promote healthy eating habits, and increase physical activity among elementary and secondary students. In support of nutrition education goals, *authorizes* technical assistance and grants to improve the quality of school meals.
- *Milk.* Places into law rules governing the offering of milk and milk substitutes that differ minimally from pre-existing law and regulations. Schools *must* offer fluid milk in a variety of fat contents, *may* offer flavored, unflavored, and lactose-free milk, and *must* provide a milk substitute for students whose "disability" restricts their diet (on written request of a physician). In addition, schools *may* substitute a non-dairy

beverage that is nutritionally equivalent to milk for students who cannot consume milk because of a "medical or other special dietary need" (on the written request of a medical authority or parent).

- *"Weighted Averages."* Reinstates a *waiver* of a requirement that schools use "weighted averages" — that measure the nutrient content of meals according to food items actually chosen by students —when doing a nutrient analysis of their school meal programs.
- *Food Safety.* Increases the number of required food safety inspections, requires state and federal audits of the inspections, and mandates that school food safety programs comply with any "hazard analysis and critical control point" (HACCP) system established by the Agriculture Department.
- *Irradiated Foods.* Requires the Agriculture Department to establish procedures relating to irradiated food products distributed by the Department to school meal programs. They must ensure that irradiated foods are provided only on request, that schools receive information about irradiation technology and its relation to safe food handling, that schools get models for providing information about irradiated foods to parents and students, that irradiated products are appropriately labeled and not commingled with other foods, and that schools are encouraged to offer alternatives.
- *Technology.* Requires state child nutrition plans to contain information as to how they may use their state administrative expense grant for information management systems that monitor the nutrient content of meals and *authorizes* funds for competitive grants to defray the cost of purchasing/upgrading technology and information systems doing this task.

Fresh Fruit and Vegetables, WIC Farmers' Markets, and Locally Produced Food

- *Free Fresh Fruit and Vegetable Program.* Requires the Agriculture Department to operate a *permanent* fresh fruit and vegetable program under which free fresh fruit and vegetables are made available to students in participating schools. This expands on a pre-existing pilot project available in five states and one Indian reservation and adds three states and two Indian reservations. In both the pilot and the expanded permanent program, the program is available in only a very small number of schools in each state or Indian reservation. Mandatory annual funding is provided for this permanent program.[10]

- *WIC Farmers' Market Program.* Reduces the state match to 30% of administrative costs (rather than 30% of total costs), increases the limit on the federal share of benefits from $20 to $30 per participant per year, and makes roadside stands eligible to participate at state option. This program provides vouchers to WIC recipients to buy fresh fruit and vegetables at participating farmers' markets.
- *Locally Produced Foods. Authorizes* funding for the Agriculture Department to encourage schools to purchase locally produced foods, including grants to defray initial infrastructure costs incurred in carrying out this policy. *Authorizes* the Agriculture Department to provide competitive matching grants and technical assistance for projects that improve access to local foods through "farm-to-cafeteria" activities, procurement from small- and medium-size farms, support for school garden programs, and farm-based nutrition education projects. *Authorizes* the Agriculture Department to provide technical assistance and grants to improve access to local foods in schools.
- *Fruit and Vegetables in the WIC Program. Authorizes* the Secretary to award grants to evaluate the feasibility of including fresh, frozen, or canned fruit and vegetables as an addition to food items in regular WIC food packages.

Summer Food Service Program

- *Simplified Summer Programs ("Lugar" Rules).* Makes permanent rules that allow summer program sponsors to receive the maximum summer program subsidy rates without providing documentation of costs and expands their application to 6 additional states (beyond 13 states and Puerto Rico covered under pre-existing law).
- *"Seamless Summer Option."* Creates, in law (as opposed to administrative policy), an option that allows schools to administer summer or school vacation food service using the rules (and subsidy rates) of regular school meal programs.
- *Pilot Projects.* Establishes and funds four pilot projects — (1) a project that, in rural parts of Pennsylvania, makes it easier for the area to qualify for "open-site" summer program status (in which all participating children are served free meals), (2) grants to increase participation in the summer program through innovative approaches to transportation in rural areas, (3) modification of eligibility criteria for residential summer camps serving low-income children in order to evaluate alternative means of judging camps' eligibility to participate in the summer program, and (4) a

project in California whereby local governments and private nonprofit organizations could receive summer program subsidy rates for up to 2 meals a day during the summer, school breaks, after-school hours, and weekends (as an alternative to after-school nutrition projects, the regular summer program and the seamless summer option noted above).

Child and Adult Care Food Program (CACFP)

- *For-Profit Day Care Centers.* Makes permanent (and nationally applicable) a pre-existing rule that allows for-profit day care centers to qualify for CACFP assistance if at least 25% of the children they serve are eligible for free or reduced-price school meals (have family income below 185% of the federal poverty income guidelines).
- *Emergency Homeless Shelters: Age Limit.* Allows CACFP subsidies to be paid for free meals/snacks served by emergency homeless shelters to all children not more than *18 years old* and to children with disabilities — up from an age limit of 12 years (15 for children of migrant workers).
- *Management Support Initiative.* Reinstates and provides mandatory funding (in FY2005 and FY2006 only) for an initiative offering training and technical assistance to aid state agencies in improving their management and oversight of the CACFP.
- *Pilot Project.* In rural areas of Nebraska, lowers the eligibility threshold applied to family day care homes wishing to qualify for "Tier I" (higher CACFP subsidy rates) status — for fiscal years FY2006 and FY2007.
- *Paperwork Reduction.* Requires the Agriculture Department to examine the feasibility of reducing paperwork associated with CACFP regulations and recordkeeping requirements for state agencies, day care homes and sponsors, and child care centers.

Special Supplemental Nutrition Program for Women, Infants, and Children (the WIC Program)

Eligibility

- *Breastfeeding Women.* Increases the potential eligibility period for breastfeeding women.
- *Physical Presence.* Expands authority for state agencies to waive the requirement that each individual seeking WIC benefits be physically present to determine eligibility to cover certain infants under eight weeks of age.

Benefits[11]

- Review of Food Packages. Requires the Agriculture Department to conduct future periodic scientific reviews of foods included in WIC food packages and change them as necessary. Also requires an update of WIC food packages within 18 months after the most current review (initiated in September 2003) is completed.
- "Rounding Up." Allows state agencies to round up to the next whole can size of infant formula when issuing WIC vouchers in order to ensure that all infants receive the full authorized nutritional benefit specified by regulation.
- Use of WIC Vouchers. Requires state agencies to permit WIC recipients to transact their vouchers at any authorized store in the state.

Cost Containment Policies

- *"WIC-Only Stores."* Substantially revises procedures aimed at containing the cost of WIC vouchers by instituting new rules state agencies must follow to control costs charged by vendors whose revenue is derived primarily (50%+) from the sale of WIC food items, effectively constraining (or ending) their participation depending on the prices they charge for WIC items.[12]
- *State Alliances.* Limits the size of alliances through which state agencies join together to solicit sole-source bids from infant formula manufacturers, in order to bolster competition in the bidding process.
- *"Primary Contract" Infant Formula.* Requires state agencies to use the "primary" infant formula they have contracted with manufacturers to supply, with all other infant formulas issued as an alternative.
- *Uncoupling Milk- and Soy-based Infant Formula Bids.* Requires that large state agencies/alliances solicit sole-source bids from infant formula manufacturers using procedures under which bids and discounts (rebates) are solicited *separately* for milk-based and soy-based formula.
- *"Cent-for-Cent" Adjustments.* Mandates a cent-for-cent decrease in rebates received by states from infant formula manufacturers if there is a decrease in the lowest national wholesale price for a full truckload of a particular formula — just as pre-existing law required cent-for-cent increases when wholesale prices increase.

WIC Vendors[13]

- *Accelerated Approval.* Requires that state agencies have procedures for accepting and processing vendor applications outside the agency's normal time-frame if there will be inadequate access to the WIC program.
- *Notification of Violations.* Directs state agencies to notify vendors of an initial violation of WIC rules if they are found to have committed a violation that requires a pattern of occurrences in order to impose a penalty.
- *List of Infant Formula Providers.* Requires state agencies to maintain lists of infant formula food wholesalers, distributors, and retailers approved to provide infant formula for the WIC program and mandates that vendors purchase infant formula from those on the state agency's list.
- *EBT Costs.* Bars the imposition of any costs related to electronic benefit transfer (EBT) systems on vendors.

Use of Funds and Administration

- *Earmarked Funding.* Requires the use of $64 million a year (or the amount of WIC funding for the prior year that has not been obligated, if less) for program infrastructure, special projects to promote breastfeeding or of regional/national significance, management information systems, and special nutrition education activities (such as breastfeeding peer counselors).
- *"Spend-forward."* Increases the limit on the proportion of the nutrition services and administration component of their WIC grant that state agencies may spend in the following fiscal year — from 1% to 3%.
- *Universal Product Codes.* Mandates that the Agriculture Department establish and maintain a national universal product code (UPC) database for states' use in carrying out the WIC program.

State Administrative Expenses

- *Minimum State Grants.* Increases the minimum state grant for child nutrition program administrative expenses from $100,000 to $200,000 (indexed after FY2008).

Technical Assistance and Training[14]

- *Food Service Management Institute (FSMI).* Adds four areas in which the FSMI is required to provide technical assistance and training — "hazard analysis and critical point plan" (HACCP) implementation by schools carrying out new food safety requirements, emergency readiness, responding to food recalls, and food bio-security training. Increases mandatory funding for the FSMI.

- *Technology and School Breakfast Expansion.* Adds two new potential uses for training and technical assistance funding — (1) aid to states (on a competitive basis) for the purpose of helping schools with large populations of low-income children in meeting the cost of acquiring/upgrading technology and information management systems used for child nutrition programs and (2) assistance to states (on a competitive basis) with low proportions of schools/children participating in the School Breakfast program in meeting costs associated with initiating/expanding the breakfast program.
- *Procurement Training. Authorizes* the Agriculture Department to provide technical assistance and training in the procurement of goods and services for child nutrition programs, including help in ensuring compliance with "Buy American" requirements and procuring safe foods (e.g., model specifications).
- *Information Clearinghouse.* Reinstates, extends, and increases the appropriations authorization for a small nongovernmental information clearinghouse that provides information to nongovernmental groups that assist low-income individuals and communities with food assistance and self-help activities to reduce reliance on government agencies.

Commodity Support

- *Permanent Authority.* Makes permanent two pre-existing commodity support authorities — (1) the requirement that the Agriculture Department use Section 32 and Commodity Credit Corporation funds to maintain annually programmed levels of commodity support for child nutrition programs and (2) permission to use Section 32 funds to remove and dispose of unsafe foods donated to child nutrition programs by the Agriculture Department.

Other Legislated Changes

In addition to the 2004 child nutrition reauthorization (and the two related short-term extensions of expiring authorities noted in the earlier History section), several relatively minor changes to the laws and policies governing child nutrition and WIC programs have been enacted in the 108th and 109th Congresses:

- P.L. 108-30. Extended the availability of funds for the project providing fresh fruit and vegetables in schools through the 2003-2004 school year.[15]
- P.L. 108-375. Included in the FY2005 National Defense Authorization Act was a provision requiring that child nutrition and WIC programs disregard "supplemental subsistence allowances" received by military families when judging eligibility.
- P.L. 108-447. Included in the Agriculture Department component of this omnibus FY2005 appropriations measure were two child nutrition/WIC provisions: (1) permission for the Department to reallocate among states unused Child and Adult Care Food Program (CACFP) audit funds and (2) a rule barring approval of any new retailers under the WIC program whose major source of revenue is derived from WIC program vouchers ("WIC-only" stores).
- P.L. 109-97. Included in this FY2006 Agriculture Department appropriations law were five child nutrition/WIC provisions: (1) continued permission to reallocate CACFP audit funds, (2) continuation of the rule barring approval of any new WIC-only stores, (3) an amendment speeding up the availability of funds to be used by the Agriculture Department to support development of local school "wellness policies" (to be available immediately instead of July 2006), (4) expansion of simplified summer program ("Lugar") rules to cover seven more states (allowing sponsors to receive maximum subsidy rates without cost documentation), and (5) provision of $6 million to expand the program offering free fresh fruit and vegetables in selected schools to six additional states.

PROPOSED LEGISLATION

Proposals Since the 2004 Reauthorization Law

A number of bills that would directly affect child nutrition programs have been introduced since the enactment of the 2004 Child Nutrition and WIC Reauthorization Act on June 30,2004.[16] The following listing does not include legislative proposals made by the Administration in its FY2007 budget submission; for details on those, refer to CRS Report RL33412, *Agriculture and Related Agencies: FY2007 Appropriations*, coordinated by Jim Monke.

- *H.R. 4800* (108th Congress). Expands the program providing free fresh fruit and vegetables in selected schools nationwide.
- *H.R. 203* (109th Congress). Expands the program providing free fresh fruit and vegetables in selected schools to one additional state (New York), and includes Head Start programs as potential participants.
- *H.R. 204* (109th Congress). Increases the amount of money for CACFP audits from 1% to 1.5% of CACFP total funding.
- *H.R. 844* (109th Congress). Establishes automatic eligibility for free school meals for children of active duty enlisted members of the Armed Forces and National Guard.[17]
- *H.R. 1589* (109th Congress). Family and Workplace Balancing Act. Along with changes to programs outside child nutrition programs, establishes a "universal" free breakfast program, directs grants to schools to create "healthy school nutrition environments," and expands the availability of CACFP subsidies for suppers served through after-school programs to *all* states.[18]
- *H.R. 5344* (109th Congress). Summer Food Service Program Improvement Act of 2006. Lowers the threshold that summer program sponsors must meet to gain "open-site" status (where all meals/snacks are served free in projects located in low-income neighborhoods). Extends applicability of "Lugar" rules to all states (see S. 1005, below). Establishes a startup grant program to provide assistance to sponsors wishing to initiate summer food service programs.
- *S. 1005* (109th Congress). Extends applicability of simplified summer program ("Lugar") rules to *all* states; these rules allow summer sponsors to receive maximum subsidy rates, without cost documentation, in 19 states and Puerto Rico.
- *S. 1695* (109th Congress). Hurricane Katrina Food Assistance Relief Act. Establishes specific authority in child nutrition law to modify child nutrition program conditions of operation in the case of natural disasters, and provides extra funding for the WIC program in the wake of Hurricanes Katrina and Rita.
- *S. 2592/H.R. 5167* (109th Congress). Child Nutrition Promotion and School Lunch Protection Act. Requires the Secretary of Agriculture to revise (and effectively expand) the definition of "foods of minimal nutritional value" that may not be offered in competition with school meals. Requires that the expanded definition bar these food items from being offered on school campuses at any time.

Proposals Pending During Consideration of the 2004 Reauthorization Law

In th 108th Congress, a number of bills were introduced to make changes in child nutrition and WIC programs — in addition to the omnibus bills reported by the House Education and the Workforce Committee and the Senate Agriculture, Nutrition, and Forestry Committee (*H.R. 3873* and *S. 2507*), which formed the basis for the final 2004 reauthorization law, and the two bills that became the laws extending child nutrition and WIC authorizations while Congress considered reauthorization (*H.R. 3232* and *S. 2241*). Many of these initiatives (or variations) were included in the House and Senate committee bills and the final law. They are as follows —

- *H.R. 1551.* Safe School Food Act. Adds to food safety requirements for schools participating in school meal programs.
- *H.R. 2227.* Obesity Prevention Act. Establishes grants for local school nutrition and physical fitness ("wellness") projects.
- *H.R. 2592.* Healthy America Act. Expands the availability of fruit and vegetables in child nutrition and WIC programs.
- *H.R. 2626/S. 1755.* Farm-to-Cafeteria Projects Act. Authorizes grants for farm-to-cafeteria projects.
- *H.R. 2832.* Healthy Nutrition for America's Children Act. Expands the program providing free fresh fruit and vegetables in selected schools nationwide.
- *H.R. 2987/S. 1392.* Establishes "healthy school nutrition environment" grants and broadens rules regulating the offering of foods in competition with school meals.
- *H.R. 3022.* School Nutrition Antibiotic Safety Act. Bars the Agriculture Department from acquiring chickens that have been fed/administered fluoroquinolone antibiotics for distribution to school meal programs.
- *H.R. 3120.* Right to Know School Nutrition Act. Mandates the provision of information about the irradiation of foods used in school meal programs.
- *H.R. 3250/S. 1367.* Child Nutrition Improvement Act. Includes provisions to increase milk consumption in schools and support for "healthy school environment" projects.
- *H.R. 3416.* Healthy Children Through Better Nutrition Act. Omnibus alternative child nutrition/WIC reauthorization proposal.

- *.R. 3869.* Pride in the Lunch Line Act. Authorizes grants to assist schools in upgrading technology and information systems used in operating school meal programs.
- *S. 448/H.R. 936.* Leave No Child Left Behind Act. Among other provisions, expands the CACFP.
- *S. 995.* Child Nutrition Initiatives Act. Provides for (1) grants to encourage more nutritious school meals and (2) grants for farm-to-cafeteria projects. Increases child nutrition state administrative expense grants for small states and funding for WIC farmers' markets. Provides funding for the Nutrition Education and Training (NET) program.
- *S. 996.* Commodity Distribution Act. Increases federally provided commodity support for school meal programs.
- *S. 1007/H.R. 4519.* Better Nutrition for School Children Act. Broadens rules regulating the offering of foods in competition with school meals.
- *S. 1020.* School Breakfast Improvement Act. Increases School Breakfast program subsidies; provides for start-up and expansion grants for breakfast programs; authorizes "universal" free breakfast pilot projects.
- *S. 1021/H.R. 3930.* Summer Food Service Improvement Act. Provides for the expansion of the Summer Food Service program.
- *S. 1022.* Child and Adult Care Food Program Improvement Act. Provides for the expansion of the CACFP.
- *S. 1393.* Expands the program providing free fresh fruit and vegetables in selected schools.
- *S. 1549/H.R. 3441.* Provides for the phase-out of reduced-price school meals by raising the income eligibility test for free school meals.
- *S. 1750/S. 1829.* Better Eating for Better Living Act. Increases school lunch subsidies; requires periodic reviews of school meal nutrition guidelines; modifies rules regulating milk served in schools; provides funding for the Nutrition Education and Training (NET) program.
- *S. 2183.* Early Attention to Nutrition Act. Establishes "Team Nutrition Network" grants to promote the nutritional health of schoolchildren.

REFERENCES

[1] In addition to these laws, Congress has used annual appropriations measures and omnibus agriculture legislation to direct the Agriculture Department to take actions affecting child nutrition and WIC programs. These directives typically take the form of requirements related to commodity support for

child nutrition efforts — but they also can encompass amendments to the underlying law or changes in program policies.

[2] For more information, see CRS Report RS20235, *Farm and Food Support under USDA's Section 32 Program*, by Geoffrey S. Becker.

[3] For information on this law and other child nutrition and WIC legislation in the 104th and 105th Congresses (1995 through 1998), see CRS Report 96-987, *Child Nutrition Legislation in the 104th Congress*, and CRS Report 97-108, *Child Nutrition Issues in the 105th Congress*, both by Joe Richardson (available on request).

[4] It should be noted that appropriations for the majority of child nutrition programs (e.g., the school meal programs) are *permanently* authorized.

[5] Virtually all child nutrition programs are mandatory entitlements. On the other hand, the WIC program is discretionary.

[6] Five authorities were made permanent: (1) a requirement to use Section 32 funds to maintain commodity support for child nutrition programs, (2) simplified summer food programs, (3) certain rules governing the treatment of for-profit child care centers, (4) the fresh fruit and vegetable program, and (5) permission to use Section 32 funds to dispose of unsafe foods. One authority (a "management support initiative" in the Child and Adult Care Food program) that expired with FY2003 was reinstated for FY2005 and FY2006. A waiver of a requirement on how schools measure the nutrient content of their meals (use of "weighted averages") that expired with FY2003 was reinstated, as was authority for an Information Clearinghouse. One expiring authority (authorization of funds for certain training and technical assistance activities) was extended through FY2008. One authority (for a nutrition education and training program) was replaced with a new initiative. And one expiring authority — a small effort funding food and nutrition projects that are integrated with elementary school curricula — was not renewed.

[7] Compliance and accountability reviews also are referred to as the "Coordinated Review Effort" (CRE). The amendment increases the authorization from $3 million to $6 million a year; however, these reviews have typically been appropriated funds above the $3 million-a-year authorization.

[8] Also, see the "Program Integrity" sections of this report.

[9] Schools receive severe-need breakfast subsidies if they served 40%+ of their lunches free or at a reduced price in the second preceding year.

[10] A later law, P.L. 109-97, added and provided new funding for six more states under this program.

[11] Also, see the section of this report entitled "Fresh Fruit and Vegetables, WIC Farmers' Markets, and Locally Produced Food."

[12] A separate provision generally bars WIC-only stores that provide "incentive items" or other free merchandise to program participants.

[13] Also see Cost Containment Policies, "WIC-only" stores, above.

[14] Also, see the "Program Integrity" sections of this report.

[15] This pilot project preceded the permanent program established in the 2004 child nutrition reauthorization law.

[16] The bills noted here do not include general "anti-hunger" measures and omnibus initiatives introduced in the 108th and 109th Congresses to reduce/prevent childhood obesity and improve schoolchildren's health. A bill in the 109th Congress that would indirectly affect school meal program operations by authorizing funding for vending machines offering healthy foods in schools (H.R. 2763, the Student Nutrition and Health Act) also is not included.

[17] This bill is the same as H.R. 5229 in the 108th Congress.

[18] These proposals also were put forward in H.R. 3780 in the 108th Congress. CACFP subsidies for suppers are currently offered in seven states.

In: Food Safety and Nutrition
Editors: J. Miller, P. Bonete, pp. 71-84
ISBN: 978-1-60456-335-1

Chapter 4

FOOD SAFETY: NATIONAL UNIFORMITY FOR FOOD ACT*

Donna V. Porter
Specialist in Life Sciences Domestic Social Policy Division

ABSTRACT

For more than a century, national uniformity of food laws has been a long-sought goal of the food industry. While early statutes tried to improve uniformity, state and local requirements established in the 1800s continued in effect throughout the United States. Food traveling in interstate commerce frequently required manufacturers to meet different nutrition labeling and/or safety standards in different states, creating additional expense and confusion.

In 1990, passage of the Nutrition Labeling and Education Act (NLEA) achieved uniformity for nutrition labeling throughout the country. Provisions in the act required that all state and local provisions that were not identical to federal standards be preempted, once a study of all federal, state, and local requirements was completed. The study was conducted to determine which requirements would be preempted and which ones were unique to states/localities and should be allowed to remain in force. The NLEA allowed states to petition FDA for an exemption from preemption, if the requirement would not be in violation of any applicable federal requirement, or unduly burden interstate commerce, and provided it addressed a particular need not met by the requirements of the law. The original bill that became

---

* Excerpted from CRS Report RL33559, dated January 23, 2007.

NLEA had contained a provision to preempt state food safety provisions, but it was dropped before passage.

Legislation has been introduced on food safety uniformity in every Congress since the 105th. In the 109th Congress, *The National Uniformity for Food Act of 2005* (H.R. 4167) would amend the Food Drug and Cosmetic Act (FDCA) to prohibit any state or other locality from establishing or continuing in effect for food traveling in interstate commerce any requirement not identical to an existing federal provision, including provisions related to adulterated food; raw agricultural commodities containing unsafe pesticides; unapproved irradiated foods; unsafe color or food additives; tolerances for poisonous ingredients; conditions for emergency permit control and their suspension; access for inspection and dietary supplement labeling regulations. The bill would allow a state to petition for an exemption or to establish a national standard related to food regulation. States would be allowed to establish requirements that otherwise would violate a FDCA provision, if the requirement is needed to address an imminent hazard that is likely to result in serious adverse health consequences. H.R. 4167 was passed by the House. *The National Uniformity for Food Act of 2006* (S. 3128) was introduced, with several changes from the House-passed version. A Senate hearing was held, but no final action was taken.

The food industry supports the bill, which it believes would improve interstate commerce by extending national uniformity to most aspects of food adulteration, preventing separate food safety notices by the 50 states and eliminating food safety warnings that are not identical with federal provisions. Opponents include food and drug officials, state attorneys general, national consumer and California groups. They believe the bill would have a serious impact on the nation's regulation of food safety and jeopardize the ability to fight bioterrorism. Opponents are also concerned about the impact of the bill on the cooperative programs for milk safety, shellfish sanitation and retail food protection. This report will be updated as action occurs.

INTRODUCTION

On March 8, 2006, the House passed the *National Uniformity for Food Act* (H.R. 4167) by a vote of 283-139. The bill would establish uniform requirements for food safety warning labels nationwide. Similar legislation (S. 3123) was introduced in the Senate, but no further legislative action was taken on either measure. Food safety labeling legislation was first considered in connection with the Nutrition Labeling and Education Act of 1990 (P.L. 101-535), but the food safety uniformity provision was dropped from the bill before passage. This report provides some background on food safety labeling, reviews past and current legislation, and discusses the potential impact of H.R. 4167 and S. 3128. It will be updated as new action is taken on future legislation.

BRIEF HISTORY OF FOOD SAFETY LABELING LEGISLATION

During the 1800s, early food law in the United States began in towns, followed by cities, counties, and states, all of which enacted requirements that initially addressed concerns about economic fraud and then food safety problems. While food was produced and consumed locally, these requirements were adequate for providing consumer protection. But as the country grew and food began to travel in interstate commerce, individual state requirements were frequently different and sometimes in conflict, and so a product in compliance with labeling and/or safety requirements in one state would not be in compliance in another state. There also was recognition that nutrient and food safety requirements generally were the same for most citizens.

In 1906, the Pure Food and Drug Law Act, the first federal statute administered by the Food and Drug Administration (FDA), defined which products were considered to be foods and regulated the adulteration and misbranding of these products. The Meat Inspection Act of 1906 granted the U.S. Department of Agriculture the power to conduct sanitary inspections in meat packing plants. Passage of the Federal Food, Drug and Cosmetic Act of 1938 (FDCA) gave FDA further authority over most food regulation, including providing a series of definitions elaborating on the concepts of adulteration and misbranding, control over all labeling of foods traveling in interstate commerce, detailed regulation of issues concerned with safety and wholesomeness of foods, and enforcement remedies available to the agency, when needed. Since 1938, FDCA has been amended numerous times to address various food safety and nutrition issues as science and regulation have evolved. Frequently, state and local requirements were the source of national standards and federal requirements that subsequently were adopted for the entire country. Nevertheless, national uniformity has not been achieved, because state and local governments have continued to pass laws to address regional needs or cover gaps in federal law.

California's Proposition 65

In 1986, Proposition 65 (the Safe Drinking Water and Toxic Enforcement Act of 1986) was approved in a California voter referendum. This consumer protection measure requires manufacturers to provide clear and reasonable warnings on products that contain chemicals listed by the State of California as reproductive toxins or carcinogens. The warnings apply to occupational exposures, ambient environmental exposures and exposures from consumer products, including foods. While foods in compliance with existing administrative standards would be expected to be unaffected by these provisions, the warnings would be required on food that contains or was exposed to chemicals known to cause cancer or birth defects at levels that pose a significant risk. Prop. 65 labeling for food requires, among other things, manufacturers to prove that their plastic packaging does not contain harmful levels of carcinogens (e.g., benzene). Elimination of Prop. 65 for food packaging would remove a barrier for manufacturers who must currently demonstrate that chemicals in the plastic do not leach into food at levels that would cause concern. The statute has had a wider impact in eliminating the use of certain hazardous chemicals altogether.

Since its adoption, Prop. 65 has been criticized by the food industry, because of its potential impact on food labeling. However, to date, there is no evidence of a problem affecting a food manufacturer who was required to provide different information on labels of the same product sold in different states as a result of the Prop. 65.[1] Nevertheless, since it passage, the food industry has sought to get Prop. 65 overturned in California, or preempted by a federal statute requiring uniform food safety labeling. It has also opposed passage of bills similar to Prop. 65 in states other than California.[2]

Nutrition Labeling and Education Act of 1990

Passage of the Nutrition Labeling and Education Act of 1990 (NLEA; P.L. 101-535) represented major food and nutrition labeling reform, following broad agreement among members of Congress, the food industry, and health and consumer groups that existing food labeling standards were outdated. The act made nutrition labeling mandatory for most foods, specified the information that was to appear, and allowed certain claims to be made. The law also required national uniformity for all nutrition labeling provisions, following a review and study of all state and local requirements. The study, which was conducted by the National Academy of Sciences' Institute of Medicine, was to determine which requirements would be preempted and which ones were unique to states/localities and should be allowed to remain in force.[3] FDA used the completed study to publish its final rule on preemption of state requirements in 1993.[4] While most state and local food labeling requirements were preempted, there were a number of state provisions that remained in force. Those included state labeling requirements unique to a specific locality (e.g., alligator meat in Florida and Louisiana), and labeling for specific food types for which FDA had yet to establish a federal standard (e.g., 23 state standards for honey). NLEA allowed states to petition FDA for an exemption from preemption, provided it would not be in violation of any applicable federal requirement or unduly burden interstate commerce, and so long as it addressed a particular need not met by the requirements of the act. At the time, many state requirements that were preempted had been on the books for decades and had never been rescinded, even though many requirements were no longer being enforced.

During legislative consideration of NLEA, food and grocery organizations voiced concern about the complex array of state and local food safety laws and regulations that required their compliance.[5] These groups argued that national uniformity for food safety requirements (both safety standards and labeling) was needed as much as nutrition labeling uniformity was needed. While national uniformity for food safety provisions was initially included in bill language, the final Act contained an exemption from national uniformity for state food safety notification labeling requirements. Uniform safety labeling requirements were seen as too controversial a provision at that time and threatened to undermine passage of the nutrition labeling reform, which was the primary focus of the legislation. As a result, the uniformity or preemption provision for food safety standards was dropped with the understanding among the interested parties in Congress and the food industry that the issue would be taken up at a later date.

National Uniformity for Food Act

Bills concerned with food safety labeling uniformity began being introduced in 1998. In the 105th Congress, the *National Uniformity for Food Act of 1998* (S. 2356 and H.R. 4383) was introduced by Senator Roberts and Representative Burr and referred to the Senate Committee on Labor and Human Resources, and the House Committee on Commerce. The measure would have preempted state and local food safety labeling requirements. In the 106th Congress, the *National Uniform Food Safety Labeling Act* (H.R. 1346) was introduced by Representative Pallone. It dealt with the addition of consumer information labels on specific food products (see discussion below). The *National Uniformity for Food Act of 1999* (H.R. 2129 and S. 1155) was reintroduced by Representative Burr and Senator Roberts in the same Congress. Both House bills were referred to the Committee on Commerce, and the Senate bill was referred to the Committee on Agriculture, Nutrition and Forestry. S. 1155, as an amendment in the nature of a substitute, was reported out of committee and placed on the Senate Legislative Calender, but no full Senate vote was taken.

Both Representatives Pallone and Burr reintroduced their bills with the same titles and virtually the same language in the 107th Congress, as H.R. 1816 and H.R. 2649, respectively. The bills were referred to the Committee on Energy and Commerce. In the 108th Congress, the bills were introduced again by the same members, as H.R. 1495 and H.R. 2699, and again referred to the Committee on Energy and Commerce. In the fall of 2003, the *National Uniformity for Food Act of 2004* (H.R. 2699), Representative Burr's bill, was reported out of committee and placed on the Union Calendar; however, the measure was never brought to the House floor for a vote. The committee report (H.Rept. 108-770) indicated that the bill had been scored by the Congressional Budget Office (CBO) to cost FDA $106 million over five years to handle an estimated 120 petitions from states. CBO was anticipating the states would request that an existing requirement in their jurisdiction not be preempted (see discussion below of petition provisions). The Senate report listed the numerous groups that opposed the bill; no list of supporters was provided.[6] No hearings on any of the bills or the issue in general have been held to date.

Legislative Activity in the 109th Congress

The *National Uniform Food Safety Labeling Act* (H.R. 2235) was introduced by Representative Pallone on May 10, 2005. The bill amends FDCA to deem food misbranded, unless the label contains certain information on specific foods. For raw or partially cooked eggs, fish, milk, dairy products and shellfish, or unpasteurized juices, labeling must disclose that eating any food raw or in a partially cooked state may pose an increased risk of foodborne illness particularly for children, the elderly, pregnant woman, and persons with weakened immune function. Frozen fish or shellfish would be required to disclose prominently that the product has been frozen, unless it is smoked, cured, cooked or commercially sterilized prior to being frozen. For perishable agricultural commodities or derivatives, labeling would need to provide the country of origin. All products would be required to disclose the date upon which they should no longer be sold because of diminished quality, nutrient availability or safety. Products would only be allowed to use the term "natural" if the food contained no artificial or synthetic ingredient added after harvesting and had not undergone anything more than minimal processing. H.R. 2235 was referred to the Committee on Energy and Commerce, but no further action has been taken.

The *National Uniformity for Food Act of 2005* (H.R. 4167) was introduced by Representative Rogers on October 27, 2005, with 226 cosponsors, and referred to the Committee on Energy and Commerce. The bill would amend FDCA to prohibit any state or other locality from establishing or continuing in effect for food traveling in interstate commerce any requirement that is not identical to an existing FDCA provision, including provisions related to: adulterated food; raw agricultural commodities containing unsafe pesticides; unapproved irradiated foods; unsafe color and food additives; tolerances for poisonous ingredients; conditions for emergency permit control and their suspension; and access for inspection regulations. The uniformity requirement would apply to warnings in food labels and labeling, advertising, posters, public notices and any other means of communication, whether adopted by statute, regulation or other administrative act. The measure would include any form of notification requirements, whether by a law specifically classified as a food statute, a consumer protection or unfair competition law, or a measure that more generally applies to all chemicals present in consumer products or the environment.

H.R. 4167 would allow a state to petition for two different reasons: for an exemption so that they could continue to use a labeling requirement that the state had established, or to request the establishment of a national standard to replace existing state requirements. The petition must identify the state statute involved, the specific food or components affected, the warning or food safety requirements imposed on that food or component, and adequate data and information to justify the public health and safety need for the requirements either as an exemption from national uniformity or a national standard. FDA would have to act within 270 days of enactment to publish a notice concerning any petition and provide 180 days for public comment on it. FDA would have a year to take final agency action after the end of the comment period. The Secretary of Health and Human Services would be allowed to provide an exemption, if the requirement that is the subject of the petition protects a public interest that is otherwise not protected, and provided that it would not cause any food to be in violation of any federal law or unduly burden interstate commerce.

H.R. 4167 would allow a state to establish requirements that would otherwise violate an FDCA provision relating to national uniform nutrition labeling or the act, if the requirement is needed to address an imminent hazard to health that is likely to result in serious adverse health consequences (e.g., contaminated food). The bill does not preempt state and local laws relating to certain labeling provisions (including freshness dating, open date labeling, grade labeling, a state inspection stamp, religious dietary labeling, organic or natural designation, returnable bottle labeling, unit pricing or statement of geographic origin), or a consumer food sanitation advisory established or recommended by the Secretary. The bill was reported out of committee on December 15, 2005,[7] and was scheduled for a House vote on the Suspension calender the week of February 27, 2006. The House Rules Committee allowed an hour of debate on the bill on March 2, 2006.[8] Following debate on several amendments (see box below), a final vote of 283-139 in favor of the bill was cast on March 8, 2006.[9] The bill is currently before the Senate.

Amendments Adopted to H.R. 4167

— Amendment to clarify that uniformity in notifications requirements for warnings does not apply to dietary supplements; and ensures that states can set tolerances for substances in food when the federal government has not.

— Amendment to provide for expedited consideration for state petitions that seek adoption of national warning requirements or exemptions from uniformity for state warning requirements in cases where the requested warning: relates to cancer-causing agents; is related to reproductive effects or birth defects; or is intended to provide information that will allow parents or guardians to understand, monitor, or limit a child's exposure to cancer-causing agents or reproductive or developmental toxins.

— Amendment to state that the changes of law made by this legislation will not take effect until after the Secretary of HHS certifies to Congress, after consultation with the Secretary of Homeland Security, that implementation of H.R. 4167 will not pose an additional risk to the public health or safety from terrorist attacks relating to the food supply.

— Amendment to prevent H.R. 4167 from affecting any state requirement that establishes a notifications requirement regarding the presence or potential effects of mercury in fish or shellfish.

Based on information from FDA and a review of state requirements likely to be affected by the bill, CBO has estimated that states would submit about 200 petitions to FDA early in 2007 and an additional 40 petitions over the 2008-2011 period. CBO further estimates that, in implementing H.R. 4167, FDA would incur total costs of less than $500,000 in 2006 and approximately $100 million over the period of 2006-2011, at an average cost to the agency of about $400,000 per petition.[10] A legal analysis prepared for the National Uniformity for Food Coalition, which represents the food industry view (see below), estimated that H.R. 4167 would only affect 11 state laws and regulations as part of the effort to set national food safety standards and warning requirements.[11] In the industry's view, the bill should be re-scored, because CBO included certain state requirements that would not be preempted by H.R. 4167. An recent analysis completed by Center for Science in the Public Interest estimated that 220 state requirements would be preempted by the act.[12]

On May 24, 2006, *the National Uniformity for Food Act of 2006* (S. 3128) was introduced by Senator Burr with two cosponsors. The legislation is very similar to the House-passed bill (H.R. 4167), but with several changes that address concerns raised by opponents. The preemption language in the Senate version is modified to clarify that states may act on issues, such as setting a tolerance level in food, when the FDA has not acted. The Senate bill would not prohibit a state from conducting inspections involving food adulteration. It verifies that FDA must take final agency action on a petition concerning an exemption. The Senate bill calls for expediting petitions for notification requirements that would enable pregnant women to understand, monitor or limit their exposures to toxins that could affect fetal development. Finally, S. 3128 does not contain the mercury amendment to the House-passed bill, at least in part as the result of the recent California ruling that rejected warning labels on canned tuna.[13] (A California state judge ruled that the state law conflicts with the federal FDA policies on tuna consumption, the mercury levels are not high enough to warrant health warnings, and tuna is exempted from warnings because mercury in fish is naturally occurring. This decision is viewed by some proponents of H.R. 4167 as a direct challenge to California's Proposition 65).

On July 27, 2006, the Senate Committee on Health, Education, Labor and Pensions held a hearing on S. 3128.[14] The industry witnesses described the benefits to the food industry of enacting a preemption statute for food safety and warnings comparable to other preemption provisions for food labeling, drugs, and pesticides. A cereal manufacturer described his experience of being sued under the provisions of Prop. 65. A former FDA deputy commissioner expressed concern that FDA did not have the resources to take on the tasks that the states have been handling for decades. However, the Senate took no final action on the bill in the 109th Congress.

Impact of the National Uniformity for Food Act

Numerous groups have lined up on either side of H.R. 4167/S. 3128. Supporters of the bill include the National Uniformity For Food Coalition, which lists 115 food companies and trade associations as members.[15] Those opposed to the legislation include certain governmental groups (e.g., the Association of Food and Drug Officials,[16] the National Association of State Departments of Agriculture,[17] the National Association of Attorneys General),[18] several national consumer groups, and various California groups.

The Coalition seeks food safety labeling uniformity to bring it in line with various other related areas of law, including food, drug, and pesticides. It believes that the legislation would improve interstate commerce by extending national uniformity to all aspects of food adulteration other than food sanitation, by preventing separate food safety notices by the 50 states, and by eliminating all food safety warnings that are not identical to federal safety warnings. National uniformity is not applied to food sanitation because states have traditionally provided a leadership role throughout the country in regulating sanitary food practices at the state and local level. Proponents of the bills have conceded that one of the primary goals is to preempt California's Proposition 65 and prevent similar requirements from being implemented in other states.

Supporters of the legislation note that all states would be allowed to petition the FDA within 180 days of enactment to exempt their state food safety laws from elimination under the bill, and state requirements would remain in effect while FDA considers the request. States could also apply to FDA to nationalize their stricter state warnings, if they had scientific evidence that some food or additive was hazardous and not addressed by federal law. Supporters believe that the bill would continue to allow a state or local entity to address emergencies under its imminent hazard authority. The food industry analysis estimated that only 11 state requirements would be affected by enactment of the legislation.

Those opposed to the bills believe that it would have a serious impact on the nation's regulation of food safety and could jeopardize the ability to fight bioterrorism.[19] Opponents have expressed concern about the lack of full debate on the impact of the legislation. Opponents argue that states are viewed as the primary guardians of food safety and that the bills would eliminate every state and local requirement that provided for greater consumer protection and preempt all existing warnings about the safety of foods.

Concern has been expressed about the interpretation of the term "identical" in the bill and the phrase "and for other purposes" in terms of preemption and the reach of the act's provisions. Opponents fear that interpretation of these terms could lead to preemption of all state and local requirements, including imminent hazard authority used to address emergencies. An imminent hazard to the public health exists when there is sufficient evidence to show that a product or practice poses a significant threat of danger to health. Such a threat creates a public health situation that should be corrected immediately to prevent injury and not be permitted to continue, while a hearing or other formal proceeding is being conducted. Many situations involving food safety that would be considered to represent an imminent hazard are identified during inspections, 80% of which currently are done by state and local inspectors.

Opponents have also raised concern about the cooperative programs for milk safety, shellfish sanitation and retail food protection, where the states carry out compliance and enforcement activities and FDA provides training, standardization, oversight and evaluation. They believe that states might lose their legal foundation under the legislation for conducting the FDA cooperative programs, if all state laws were preempted. FDA operates these cooperative programs by providing model ordinances or codes (which are not regulations that have undergone notice and comment) that states adopt into their laws and that rely on state funds, licences and permit fees to support the operation of the programs. Potentially these operations and the source of funding would be eliminated by the legislation's passage, requiring FDA to promulgate regulations, that could take years to finalize. The 2003 House report estimated that 80 state laws in 37 states would be eliminated by the bill's passage.[20] The current CBO scoring of H.R. 4167 only addresses the estimated costs to review petitions. It does not address the costs of setting up programs that would replace the state's current efforts under their existing laws. The recent estimate by Center for Science in the Public Interest is that 220 state requirements would be affected.

Finally, opponents, particularly those from California, believe that the bills would prevent states from imposing warnings against cancer or other health problems, where FDA has not acted or when the FDA's standards are weaker than a state believes is necessary to protect its citizens.[21] While the petition process is available, many state officials view that process as very labor intensive and costly for states who choose to use it. The legislation is unclear on whether a state's warning label would be eliminated, if the federal government has no similar warning. While Prop. 65 has resulted in some warnings, its California supporters believe it has also created a market incentive to remove dangerous chemicals from foods and to bring safer foods to market.[22]

Several questions about this legislation remain to be answered. The significant difference in the estimates by CBO, CSPI, and the food industry of the number of state requirements that potentially would be affected by the bills is indicative of the different perspectives about the legislation's reach. Proponents claim that the impact of the legislation would be limited, because the bills would not preempt state laws on food sanitation, inspections, enforcement, nutrition standards, response to imminent hazards, state requirements identical to federal requirements, notification requirements (unless requiring warnings about food safety) or the safety of dietary supplements. Opponents believe that the bills' reach would be broad and would undermine the existing food safety system and it's protections, including the cooperative inspection programs, addressing imminent hazards and bioterrorism, and would require significant state resources to maintain the current level of protection. Another issue concerns the goal of national uniformity. The bills would achieve national uniformity by preempting certain food safety provisions, including Prop.65. However, the number of state laws that would not be preempted raises the question of the degree to which national uniformity actually will be achieved by passage of the legislation. Both sides sought the July 2006 hearing on the legislation with the hope that it would provide the opportunity to help sort out the divergent views and provide a clearer assessment of the legislation's impact on both the issues of reach and degree of uniformity.

No final action was taken in the Senate to pass the bill before the end of the 109th Congress. Since this issue continues to be of considerable concern to the food industry and state and federal regulators, bills are likely to be reintroduced in the 110th Congress. New legislation may provide an opportunity for proponents and opponents to previous bills to reach agreement on legislative provisions and achieve passage in the 110th Congress.

REFERENCES

[1] Letter to Member of Congress on H.R. 4167, the National Uniformity for Food Act, from the National Association of Attorneys General. Mar. 1, 2006, p. 5.

[2] Testimony of Susan Connelly, Manager, State Affairs Grocery Manufacturers of American Inc. Letter in Opposition to Connecticut Clone Bill Mar. 15, 2001, at [http://www.gmabrands.com/news/ docs/Testimony. cfm].

[3] Institute of Medicine, 1992, *Food Labeling: Toward National Uniformity.* Committee on State Food Labeling, Food and Nutrition Board, National Academy Press, p. 239.

[4] U.S. Dept. of Health and Human Services. Food and Drug Admin, Food Labeling; General Provisions; Nutrition Labeling; Label Format; Nutrient Content Claims; Health Claims; Ingredient Labeling; State and Local Requirements and Exemptions; Final Rules. Federal Register, v. 58, no. 3, Jan. 6, 1993, pp. 2065-2941.

[5] Institute of Medicine. 1992. *Food Labeling: Toward National Uniformity*, Committee on State Food Labeling, Food and Nutrition Board, National Academy Press, p. 54.

[6] H. Rept 108-770, National Uniformity for Food Act of 2004, p. 25.

[7] H.Rept. 109-379 on the National Uniformity For Food Act of 2005.

[8] Congressional Record-House. March 2, 2006. H530-539.

[9] Congressional Record-House. March 8, 2006. H727-758.

[10] Congressional Budget Office. Cost Estimate. H.R. 4167 National Uniformity for Food Act of 2005, Feb. 27, 2006, p. 5.

[11] Coalition Analysis Sets Record Straight on Impact of Uniformity for Food Act; Eleven -Not 200- State Laws Affected at [http://www.gmabrands.com/news/docs/NewsRelease.cfm].

[12] [http://www.cspinet.org/new/200607261.html].

[13] California Judge's Ruling Scraps Senate Food Labeling Language, CQ Today, May 25, 2006.

[14] [http://help.senate.gov/Hearings/2006_07_27/2006_07_27.html].

[15] National Uniformity for Food Coalition at [http://www.uniformity forfood.org/aboutthe coalition.htm].

[16] Association of Food and Drug Officials at [http://www.afdo.org].

[17] National Association of State Departments of Agriculture at [http://www2.nasda.org/ NASDA].

[18] National Association of Attorneys General at [http://www.naag.org].

[19] Open Letter from Marion Alder, President of the Association of Food and Drug Officials to Members of the House of Representatives, Jan. 16, 2006 via Fax Transmission, p. 2.

[20] H.Rept. 108-770 suggested that state laws in the following areas would be eliminated by passage of the act: warning labels on shellfish; regulation of smoked fish; setting tolerances for adulterated pecans and other nuts; labeling requirements for the source of certain fish; labeling requirements for citrus fruits and juices; regulation of packing fish in casks, disclosure of whether uncooked fish and shellfish have been frozen and regulation of the labeling of packages of apples; laws allowing states to adopt tolerances for food and color additives that are more protective of human health than federal tolerances; laws imposing additional requirements for egg safety;

and requirements for disclosure of the presence of specific toxic chemicals in foods.

[21] Myths and truth about H.R. 4167, *National Uniformity for Food Act*, at [http://www.consumersunion.org/pub/core_food_safety/003230.html].

[22] Zhang, Jane. *Food industry advances in labeling fight*. The Wall Street Journal, Jan. 9, 2006, p. A4.

In: Food Safety and Nutrition
Editors: J. Miller, P. Bonete, pp. 85-94
ISBN: 978-1-60456-335-1

Chapter 5

REGULATION OF PLANT-BASED PHARMACEUTICALS *

Geoffrey S. Becker
Resources, Science, and Industry Division Donna Vogt Domestic
Social Policy Division

ABSTRACT

Among many genetically engineered crops being developed are a number modified to produce pharmaceutical compounds (none yet commercialized). Proponents say such plant-based proteins could be produced more cheaply than by conventional methods and economically benefit farmers. But critics see risks, such as potential environmental, food safety, and trade problems. Among the issues is whether the current system for regulating biotechnology is adequate. This report will be updated if events warrant.

INTRODUCTION[1]

Since the first genetically engineered (GE) crop varieties became commercially available in the mid-1990s, U.S. farmers have been rapidly adopting them in efforts to lower production costs and raise crop yields. Among

---

* Excerpted from CRS Report RS21418, dated March 8, 2005.

many new GE crops now in development, but not yet commercialized, are plants that could be "factories" for pharmaceutical compounds, which could be extracted and purified for specific human and animal health uses. Worldwide, hundreds of GE, plant-based pharmaceuticals are being developed. Proponents believe they could produce proteins that might, for example, be used to vaccinate against and/or treat various forms of cancer, infectious diseases, nervous system and cardiovascular diseases, metabolic disorders, and agents of biowarfare. A number of applications could be approved within 10 years, according to some experts.

In the United States in 2004, seven companies and universities had USDA permits to field test pharmaceutical production in the following crops: corn (3 permits); tobacco, or tobacco mosaic virus (TMV) using tobacco as the host (3); and safflower (1). Test plot locations for these crops were approved for Florida, Arizona, Nebraska, Kentucky, Missouri, South Carolina, Texas, and Colorado.[2]

Potential Benefits

Proponents contend that plant-based pharmaceuticals could be a far cheaper alternative — as low as pennies per dose compared with many dollars to produce those now made by more conventional methods. These conventional methods require major investments in both large volumes of purified culture mediums and in manufacturing plants. Plant-based pharmaceuticals, on the other hand, are highly cost-efficient protein producers, which could be easily incorporated into the existing agricultural infrastructure and by well-managed farms, proponents argue. Many thousands of acres could be converted to "pharming" for higher earnings. Corn farmers, for example, might gross thousands of dollars per acre for pharmaceutical varieties compared with several hundred dollars for the feed corn they grow now, advocates argue, noting that there has been support from some major corn growing regions for this technology. Many experts consider pharmaceuticals from plants to be safer than those from mammalian sources, as small amounts of pathogenic material from a host animal could mix with and contaminate the new compound. No known plant viruses have infected humans, unlike animal pathogens.

Potential Risks

Corn is the crop most extensively used and researched for GE uses, including pharmaceuticals. Corn also is the most widely planted U.S. crop at approximately

80 million acres annually, most of it destined for animal feed (and indirectly meat and poultry), and corn meal, oil and syrup used in many processed foods. Some critics want "pharmed" corn to be banned from areas where food and feed corn are produced. Critics worry that corn's open pollination makes it difficult to prevent pollen drift and gene flow into other corn varieties, and are concerned about impacts for food crops and wildlife. Also, some major foreign buyers already are reluctant to accept any U.S. crops for which GE varieties are used — despite U.S. assertions that approved GE crops (all of them non-"pharm") are no different than, and as safe to consume as, those produced by more traditional means. Keeping GE "pharmed" corn separate from food and feed implies higher production costs. Large scale production of "pharmed" corn also might bring more foreign opposition to products of agricultural biotechnology.

Regulation

The basic federal guidance for regulating all biotechnology products, including plant-based pharmaceuticals, is the Coordinated Framework for Regulation of Biotechnology (51 *Fed. Reg.* 23302) published in 1986 by the White House Office of Science and Technology Policy (OSTP). A key principle is that genetically engineered products should continue to be regulated according to their characteristics and unique features, and not according to their production method. The Framework provides a regulatory approach intended to ensure the safety of biotechnology research and products, using existing statutory authority and previous agency experience with traditional breeding techniques.

The three lead agencies are USDA's Animal and Plant Health Inspection Service (APHIS), which regulates the importation, interstate movement, and field testing of GE plants and organisms that are or might be plant pests under the Plant Protection Act (PPA; 7 U.S.C. §7701 *et seq.*) and which regulates animal biologics (i.e., viruses, serums, toxins for animal vaccines) under the Virus, Serum, and Toxins Act (21 U.S.C. 151 *et seq.)*; the Food and Drug Administration (FDA), which regulates food, animal feed additives, human and animal drugs and medical devices, including those from biotechnology, primarily to ensure that they are effective and pose no human health risks, mainly under the Federal Food, Drug and Cosmetic Act (FFDCA; 21 U.S.C. §301 *et seq.*) and the Public Health Service Act (42 U.S.C. §201 *et seq.*); and the Environmental Protection Agency (EPA), which must approve the use of all pesticides, including those genetically engineered into plants. Developers of plant-based pharmaceuticals mainly work with APHIS and FDA, often at the same time; both agencies share information about a product during its development and testing. Such products also may be subject to state laws.[3]

APHIS[4]

GE plants that are or might be plant pests are considered "regulated articles" under APHIS regulations (7 CFR 340-340.9). APHIS authorization must be obtained prior to import, interstate movement, or environmental release, including field testing. To gain "non-regulated status" (so that a plant can be freely commercialized with no further oversight), a developer must provide APHIS with extensive information on plant biology and genetics, and potential environmental and plant pest impacts that may result from the modification. APHIS conducts a formal environmental assessment (EA) and has public comment periods before deciding whether to approve the petition. It is only after years of field testing, scientific review of extensive data, and a public comment period that "non-regulated status" may be granted. Until then, a "regulated" plant cannot be introduced into the environment, even field tested, unless its developer obtains authorization through either the (1) notification or (2) permit process. Sponsors follow APHIS guidance on testing and movements to ensure that the plant will not damage agriculture, human health, or the environment. "Pharm" plant developers must utilize the permit, not the notification, process.

Notification

The majority of GE crops are developed under the notification option, a less rigorous process than permitting. Developers must obtain authorization from APHIS before interstate movement, import, or field testing of a GE organism. APHIS then either authorizes or denies the notification, within 10 days for interstate movement, and 30 days for a field test notification. Authorization is valid for one year, with renewals through additional notifications. The developer must submit a report within six months after completion of field tests and also inform APHIS of any unusual occurrence during tests. Under the notification process, the agency generally inspects about 10% of current trials annually. Once "non-regulated," a crop can go into commercial use and is no longer subject to confinement measures and other APHIS oversight.

Permit

GE plants for pharmaceuticals fall into a distinct regulatory category handled differently by APHIS than GE plants for food or feed use. The permit process is much more stringent than notification, starting with application, which must be at least 120 days in advance of environmental release, and at least 60 days in advance for movement. Required permit information is more extensive than under notification, as is the subsequent list of permit conditions. An environmental analysis (EA) may have to be conducted under the National Environmental Policy Act (NEPA; 7 CFR 372 *et seq.*), with opportunity for public comment, prior to permitting. Generally, confined field trials do not require an EA if they fit certain categorical exclusions under NEPA (7 CFR 372.5), which APHIS has applied to virtually all recent plant-based "pharm" tests. However, an EA is required when the organism involves a new species, or the organism or a novel modification raises new issues.[5] Once APHIS initially agrees to a permit, a letter and proposed conditions go to the state, which can add its own conditions before approval, or even deny the permit. APHIS says it defers to the states on final decisions about permits.

APHIS specifies field test, movement and storage procedures to ensure that GE plants and seeds for pharmaceuticals do not enter the food and feed supply; requires a chain of custody and identity preservation system that tracks seeds from shipping through planting, harvest, and product extraction; and sets forth federal auditing requirements, among other rules. APHIS inspects all pharmaceutical field trials at least annually. APHIS materials (footnote 4) summarize in more detail the confinement measures for pharmaceutical organisms that were field tested in 2002. Examples of such measures, which are crop-specific, include geographically isolating test plots; planting buffer/border rows of non GE crops; preventing pollination or otherwise blocking reproduction; adequately identifying and segregating the organism; post harvest restrictions on test plots; destruction of leftover GE plant material; and a contingency plan to deal with any accidental releases. Developers are subject to civil or possibly criminal penalties for violating permit conditions.

Unlike plants engineered for use as food or feed which may be granted non-regulated status, no GE pharmaceutical plant has been declared "non-regulated" by APHIS. Agency officials told CRS that they cannot envision deregulating pharmaceutical plants — thus meaning that in commercialization they will be grown under both APHIS permit and FDA regulation (in the case of a plant-made veterinary biologic, also APHIS).

FDA

FDA currently regulates modified plant products as it regulates human and animal drugs and human biologic products. (FDA's rules are found in 21 CFR 201 *et seq.*; 301 *et seq.*; 501 *et seq.*; 601 *et seq.*). A "pharmed" ingredient intended for therapeutic use must meet the same quality standards of safety, purity, potency, and effectiveness as traditionally synthesized pharmaceuticals. Regardless of the production method, FDA regulations require manufacturers or sponsors to file New Drug Applications (NDAs), or Biologics License Applications (BLAs) which include all reports of clinical studies and information on the safety and supply of raw ingredients used to make the finished product. Also, the ingredients must be reproducible with consistent quality over an extended time.

When a pharmaceutical company or sponsor already holds an approved application, and wants to substitute a plant-based pharmaceutical ingredient in the drug or biologic, it must file a supplemental application reconfirming the safety and efficacy of the final drug/biologic product. However, as the application is being reviewed and evaluated, the agency must reinspect the facility and production line where the new drug or biologic will be manufactured to ensure that it will be made to specification, and according to good manufacturing practices (GMPs). The entire manufacturing process, from seed banks, field preparation, planting, cultivating, harvesting, storage, transportation, extraction, to the purification of ingredients to be used in the finished product, must also comply with GMPs. The sponsor also needs to demonstrate that the pharmaceutical ingredient contains no allergens, anti-nutrients, toxic compounds, or heavy metals or other materials from the host plant that could adversely affect the safety or effectiveness of the pharmaceutical.

Under NEPA, FDA must consider the potential environmental impact of its decisions. Most drugs and biologics are eligible for exclusion from this rule because their ingredients are manufactured in enclosed facilities. Many plant-made pharmaceuticals likely would be field produced and not necessarily be excluded; however, it is likely that FDA would defer to APHIS for environmental reviews of pharmaceutical plants grown outdoors. FDA reviews for NEPA would probably focus on potential environmental impacts posed by uses after harvest that are not subject to APHIS oversight.

FDA is evaluating a number of regulatory options should plant-based GE pharmaceutical ingredients end up in the food supply. Should this happen, FDA might consider the food adulterated, and take some sort of administrative action. Some critics of current rules have suggested the agency approve plant-made pharmaceutical proteins as if they were food additives. In such instances, if trace amounts were found in food, the food would not be considered adulterated automatically, and there might not be need for a recall.[6] However, the food additive approval process is lengthy and costly. In addition, by establishing acceptable levels, the agency is stating that this material is safe for consumers, which may not be true. For these reasons, agency officials do not want to spend limited resources establishing acceptable levels when the pharmaceutical ingredients are not ordinarily found in food, an FDA official told CRS (personal communication, January 17, 2003). He added that because of consumer reaction, food manufacturers are unlikely to accept such material.

Regulatory Concerns and Issues

The ProdiGene Case

In 2002, material from GE-altered corn plants that had been test-planted in a prior growing season in Nebraska for pharmaceutical use (for ProdiGene, Inc.) was inadvertently mixed with some 500,000 bushels of soybeans, which had to be quarantined by USDA to keep them out of the food supply. In a December 6, 2002, press release, USDA announced that ProdiGene had agreed to pay a civil penalty of $250,000, to reimburse USDA for destroying the beans, post a $1 million bond, and meet higher field testing compliance standards. USDA officials observed that the soybeans never reached the food or feed supply, evidence that current regulatory oversight is effective. Critics countered that the ProdiGene case illustrates the dangers of growing plant-based drugs, and predicted a consumer backlash if government regulation is not strengthened. Proponents argue that careful management and oversight of test sites can address such concerns.

Nonetheless, concerns persist among both consumer groups and the food manufacturing industry about producing GE plant-made pharmaceuticals in food crops. These critics believe that any cross-contamination of the food supply could create a situation in which FDA would consider such foods "adulterated" under safe food law and subject to enforcement actions. Some food industry representatives want 100% prevention systems in place before the first product is commercialized. Some critics suggest that only non-food crops should be used for GE plant-made pharmaceuticals, or, at minimum, that GE food plants shed no pollen. Other potential issues include whether manufacturers of plant-based pharmaceuticals will be able to maintain consistency in dosages and overall quality; and unanticipated environmental problems (e.g., threatening endangered species).

Regulatory Modifications

Even before the ProdiGene incident, regulators were reassessing oversight procedures for GE crops in general and plant-based pharmaceuticals in particular. Also, APHIS published in the March 10, 2003, *Federal Register* a notice tightening permit conditions for any 2003 field tests of GE plants with pharmaceutical and industrial traits. The new changes include: (1) no traditional corn can be grown within one mile of pharmaceutical or industrial field test sites involving open-pollinated corn and within one-half mile of test sites for controlled pollinated corn (twice the prior distance rules), and border rows of nontransgenic corn no longer can be used to reduce these distances; (2) for all pharmaceutical crops (corn and other), fallow zones around test sites are doubled to 50 feet; (3) restrictions on what can be grown on a test site and fallow zone in the next growing season; (4) use of dedicated machinery (e.g., harvesters, planters) and storage facilities (for both the product and machinery) to be used only for pharmaceutical production — adequate cleaning for other uses is no longer acceptable; (5) equipment cleaning and seed cleaning and drying procedures must be submitted for APHIS approval; (6) an APHIS-approved staff and cooperator training program; (7) an increase in APHIS field site inspections from one per season, to five per season plus two visits the following year to look for any volunteer plants; (8) more recordkeeping requirements.

In the January 23, 2004, *Federal Register*, APHIS published a notice of its intent to prepare an environmental impact statement (EIS) evaluating its biotechnology regulations generally, and requesting public comment on a number of possible changes. These include whether to broaden APHIS's regulatory scope to cover GE plants that may pose a noxious weed risk or may be used as biological control agents; whether to establish new categories for field testing that delineate requirements based upon relative levels of potential risk; and whether to change (i.e., strengthen) its environmental reviews and permit conditions for GE plants producing pharmaceuticals and industrials. The agency received over 3,000 comments on its proposal. APHIS was expected to publish a draft EIS sometime in 2005.

Another focal point has been a draft document, published September 6, 2002, *Guidance for Industry: Drugs, Biologics, and Medical Devices Derived from Bioengineered Plants for Use in Humans and Animals* (available at [http://www.fda.gov/ cber/gdlns/bioplant.pdf]). Prepared by FDA and USDA, it represents federal agencies' recent thinking on what kinds of information sponsors should develop and submit with requests for approvals of drugs and biologics made with plant-based pharmaceuticals.

REFERENCES

[1] This section's sources include the Biotechnology Industry Organization, Pew Initiative on Food and Biotechnology, and CRS Issue Brief IB10131, *Agricultural Biotechnology: Overview and Selected Issues*. Contacts: Tadlock Cowan (7-7600) on agricultural and USDA aspects of this report; Donna Vogt (7-7285) on medical, food safety, and Food and Drug Administration aspects.

[2] Personal communication with Legislative and Public Affairs Office, USDA's Animal and PlantHealth Inspection Service (APHIS), March 7, 2005. In 2002, more than 15 companies and universities had USDA permits to field test pharmaceutical production in the following crops: corn (41 permits); tobacco, or tobacco mosaic virus (TMV) using tobacco as the host (18); rice (8); and alfalfa, barley, rapeseed, sugarcane, and tomato (1 each), according to USDA/APHIS data compiled by Information Systems for Biotechnology at Virginia Tech ([http://www.isb. vt.edu/]).

[3] *CEQ and OSTP Assessment: Case Studies of Environmental Regulations for Biotechnology* (January 2001), by the White House Council on Environmental Quality, provides a more detailed look at how the federal regulatory system for biotechnology functions, applying it to specific products, including a plant-based pharmaceutical (see [http://www.ostp.gov/ html/012201.html]).

[4] Sources include CRS Report RL30198, Food Biotechnology in the United States: Science, Regulation, and Issues; the Pew Initiative's Guide to U.S. Regulation of Genetically Modified Food and Agricultural Biotechnology Products; BIO's Reference Document for Confinement and Development of Plant-Made Pharmaceuticals in the United States; Belson, Neil A., "U.S. Regulation of Agricultural Biotechnology: An Overview," in AgBioForum; and various APHIS materials summarizing confinement and related measures for plant pharmaceutical field tests (APHIS website: [http://www.aphis.usda.gov/brs/]).

[5] APHIS said it has strengthened its EA guidelines, including posting EAs for pharmaceutical and industrial crop field tests on its website and accepting public comments on them.

[6] In 2001, StarLink™ corn, a GE corn approved only for animal and not human food, was found in taco shells. CRS Report RS20732, *StarLink™ Corn Controversy: Background.*

In: Food Safety and Nutrition
Editors: J. Miller, P. Bonete, pp. 95-104 ISBN: 978-1-60456-335-1

Chapter 6

BSE ("MAD COW DISEASE"): A BRIEF OVERVIEW*

Geoffrey S. Becker
Specialist in Agricultural Policy Resources,
Science, and Industry Division

ABSTRACT

The appearance of BSE (bovine spongiform encephalopathy or "mad cow disease") in North America in 2003 raised meat safety concerns and disrupted trade for cattle and beef producers. A major issue for Congress has been how to rebuild foreign markets for U.S. beef. Other issues include whether additional measures are needed to further protect the public and cattle herd, and concerns over the relative costs and benefits of such measures for consumers, taxpayers and industry. This report will not be updated.[1]

WHAT IS BSE?

BSE (bovine spongiform encephalopathy or "mad cow disease") is a fatal neurological disease of cattle, believed to be transmitted mainly by feeding infected cattle parts back to cattle. More than 187,000 cases have been reported

* Excerpted from CRS Report RS22345, dated December 19, 2006.

worldwide, 183,000 of them in the United Kingdom (UK) where BSE was first identified in 1986. The annual number of new cases has declined steeply since 1992. Humans who eat contaminated beef are believed susceptible to a rare but fatal brain wasting disease, variant Creutzfeldt-Jakob disease (vCJD). About 160 people have been diagnosed with vCJD since 1986, most in the UK and none linked to any Canadian or U.S. meat consumption.

BSE IN NORTH AMERICA

BSE has been reported in 11 North American-born cattle. (A 12th case was a U.K. import into Canada.) The first native case was an Alberta, Canada, beef cow reported in May 2003. Seven more cases have been found on Canadian soil, five of them in 2006, most recently in August 2006. The first U.S. case was in a Canadian-born dairy cow found in Washington state in December 2003. The other two U.S. cases were a 12-year-old Texas-born and -raised beef cow, found in November 2004 but not confirmed until June 2005, and a 10-year-old Alabama beef cow found in late February 2006.

In epidemiological investigations of the three U.S. cases, the U.S. Department of Agriculture (USDA) was unable to track down all related animals of interest, but those that were located tested negative for the disease. Despite a beef recall, some meat from the first U.S. BSE cow may have been consumed, USDA said, adding, however, that the highest-risk tissues never entered the food supply. No materials from the other two U.S. cows entered the food supply, USDA also said. In the recent Alabama case, authorities were unable to determine the cow's herd of origin. Animal health officials initially indicated that all of the North American cases were caused by the consumption of BSE-contaminated feed. However, USDA reportedly now believes that the two native-born U.S. cattle had "atypical" BSE, which differs from other cases. If these cases are determined to be "spontaneous," that may affect future control strategies.

USDA had asked an expert international review team (IRT) to assess its response to the first (December 2003) U.S. case, and to comment on the adequacy of existing BSE protections. The IRT generally concluded (in February 2003) that the investigation had conformed to internationally accepted scientific standards, but urged additional actions, including an intensive surveillance program to measure the magnitude of the BSE problem in North America. The IRT had concluded that other infected animals probably were imported here, some of their parts rendered into ingredients fed to cattle, and amplified within the cattle herd, indigenously infecting some of them.

While agreeing with some of the conclusions, USDA also responded that the IRT's recommendations were based on the premise of a higher incidence of BSE in the United States than was indicated by current studies. USDA officials have continued to rely heavily on the findings in a detailed quantitative analysis (using computer simulation) conducted for them by the Harvard Center for Risk Analysis and Tuskegee Center for Computational Epidemiology. This work concluded that BSE was very unlikely to become established or spread in the United States.

U.S. Safeguards

Import Restrictions

USDA's Animal and Plant Health Inspection Service (APHIS) is charged with protecting U.S. agriculture from foreign pests and diseases. In 1989, APHIS began to implement increasingly restrictive import controls on ruminants and associated meat products from countries where BSE has been found. In 2003, the agency refocused its policy to allow imports of lower risk products from BSE countries, as long as they could show they had effective BSE controls (Canada was first to qualify). This approach parallels new BSE guidance supported by the United States and adopted in May 2005 by the World Animal Health Organization (OIE for its French acronym).

Testing and Surveillance

APHIS has tested cattle deemed of highest risk for BSE — for example, those that display suspicious neurological symptoms, that are nonambulatory, or that die on farms. This program tested about 20,000 cattle in each of FY2002 and FY2003, out of about 35 million slaughtered annually. Following the first U.S. case, USDA in June 2004 launched an enhanced surveillance program to determine the extent of BSE in as many higher-risk cattle as possible. After more than two years, through August 2006, nearly 788,000 of these cattle had been tested (about 5,000 weekly on average). All but those cited above were negative for BSE. Also as part of its enhanced surveillance, APHIS tested more than 21,000 clinically normal adult animals in late 2005. By late 2006, USDA had ended enhanced testing and transitioned to an ongoing program to test approximately 40,000 cattle yearly.

Meat Safety

On January 12, 2004, USDA's Food Safety and Inspection Service (FSIS), responsible for inspecting all live cattle and beef products destined for human food, published several meat plant regulatory changes. These included declaring, as "specified risk material" (SRM, which are tissues where the BSE agent can accumulate): brains, spinal cords, and other nerve tissues from cattle 30 months of age and older and the tonsils and distal ileum (part of the small intestine) of all cattle — effectively banning such tissues from the meat supply. FSIS also prohibited the slaughter of "downer" (nonambulatory) cattle; and restricted certain meat plant mechanical procedures that could spread BSE infective material into meat products.

Feed Restrictions

BSE is thought to have first spread and magnified by feeding rendered by-products of infected cattle to other cattle as a protein supplement. Since August 7, 1997, the Food and Drug Administration (FDA) has banned most mammalian (including cattle) proteins from cattle feed. FDA also registers and monitors renderers, feed mills, pet food manufacturers, and others. In January 2004, FDA promised to tighten this rule by also banning poultry litter, plate waste, and bovine blood from cattle feed.

Instead, on October 6, 2005, FDA published a proposed rule banning some SRM (mainly brains and spinal cords from cattle 30 months of age and older, and from all cattle not passed for human food) from all animal feeds, including pet food. The agency said its rule would remove those cattle parts responsible for 90% of potential BSE infectivity. The public comment period ended on December 20, 2005; a final rule was pending in late 2006.

Views on Safeguards

Both the adequacy and enforcement of these government safeguards have been criticized. For example, the Government Accountability Office (GAO) has reported weaknesses in enforcement of the current FDA feed rule. A number of consumer groups and others have asserted that FDA should have been moving much more quickly to implement stricter animal feed rules than now proposed. Separately, USDA's Inspector General (IG) has questioned both the sampling and reporting aspects of the Department's BSE surveillance program. USDA's efforts to reopen the U.S. border to Canadian beef also have been sharply criticized; see below.

Others, including many cattle and meat industry leaders, generally have been supportive of the USDA and FDA safeguards, which the Administration has argued meet and often exceed OIE guidelines. At the same time, some of these same cattle and meat industry leaders have expressed concern regarding the need for, and costs of, some of the newer safeguards, such as the USDA ban on all "downer" cattle regardless of the reason for their inability to stand or walk, and the FDA proposal to expand animal feed controls.

INDUSTRY AND TRADE IMPACTS

Exports

Cattle production is the largest single segment of U.S. agriculture, accounting for 20% of the value of U.S. farm sales. Exports of U.S. beef and other cattle products are viewed as critical to long-term market growth. The value of beef and beef variety meat exports was estimated by USDA to exceed $3 billion in 2003 (or about 10% of the farm value of cattle/calves). Four countries bought approximately 90% of these exports: Japan (37%), South Korea (24%), Mexico (20%), and Canada (10%).

Most countries halted imports of U.S. beef and cattle soon after the December 23, 2003, U.S. BSE announcement. Many but not all of these countries are again accepting at least some U.S. beef and other products. USDA reports that U.S. beef and veal exports globally reached 461 million pounds in 2004 and 689 million pounds in 2005, compared with a 2003 level of 2.523 billion pounds. The U.S. share of the world beef/veal export market declined from 18% in 2003 to about 3% in 2005. The 2006 export estimate was approximately 1.1-1.2 billion pounds and 7% of world market share.

Cattle Prices

Domestic cattle and beef prices had reached record highs in 2003 due to a tight supply-demand situation. Immediately after the first U.S. BSE case in December 2003, these prices dropped sharply, but then recovered substantially. A decline in U.S. cattle supplies, due in part to widespread drought conditions in cattle country along with strong domestic demand for beef, kept farm prices relatively high during much of 2004 and into 2005. USDA has reported that annual average U.S. prices for fed steers (i.e., slaughter-ready cattle) were $84.75 in 2004, near the lower end of a USDA forecast of $84-$91 per cwt. (100 pounds) that had been made just before the BSE case. The 2005 price was $87.28, and the 2006 estimate was $85.

In a 2005 study of the impact of the BSE situation, Kansas State University estimated that total U.S. beef industry losses due to the loss of beef and offal exports in 2004 ranged from $3.2 billion to $4.7 billion. The National Cattlemen's Beef Association earlier in 2005 had placed cattle producers' export-related losses at $175 per head.

Japan Situation

In October 2004 the United States and Japan had announced a framework for restarting beef trade. Among other things, Japan (where more than two dozen BSE cases have been found) promised to ease its domestic policy of universal BSE testing, and to admit lower risk U.S. beef if the United States could verify that it had come from cattle under 21 months old and that all SRM from the cattle, regardless of age, had been removed at slaughter. However, the market did not reopen until December 2005, when the Japanese finalized their decision to permit such U.S. imports.

On January 20, 2006, the Japanese again halted all U.S. beef imports after finding vertebral column bones (a prohibited material) in several boxes of veal shipped by a New York processor. Despite U.S. apologies and promises of stronger oversight measures, Japan did not reopen its market until July 27, 2006. Japanese safety inspections of U.S.-certified beef plants were among a number of new concessions made by the United States. Meanwhile, Japanese consumers have been substituting other proteins and other beef sources (notably, Australia and New Zealand) for U.S. beef, which once accounted for 25%-30% of Japanese beef consumption.

Korea Situation

Korea's prohibition on U.S. beef, which had been in place since December 2003, was lifted on September 11, 2006. Resumption of U.S. beef exports to Korea, the United States' second largest export destination for beef in 2003, was expected to proceed slowly for the same reasons that will slow Japan's resumption of beef imports. Strict quarantine inspection requirements in Korean ports have already resulted in the rejection of three shipments of U.S. beef because of the presence of bone fragments.

Canada Situation

USDA banned all imports of Canadian cattle, beef, and other ruminants and products in May 2003. In August 2003, USDA announced (without publishing a formal rule) that it would issue permits to import some lower-risk Canadian products, notably boneless beef. USDA soon began expanding the types of permitted beef imports in late 2003 and early 2004. In response to a Montana cattle group's lawsuit, a federal judge temporarily blocked this effort, largely on the grounds that USDA was expanding product eligibility outside of the prescribed rulemaking process.

On January 4, 2005, USDA published a formal final rule to reopen the border to additional Canadian products, including cattle 30 months and younger. The judge again blocked implementation, but a federal appeals court overruled this decision in July 2005. Since then, Canadian cattle have been entering the country. In 2006 USDA was preparing a proposed rule that also would permit imports of Canadian cattle over 30 months of age.

In a February 2005 report, USDA's IG had earlier concluded that the Department's actions on Canada were sometimes arbitrary and undocumented, policy decisions were poorly communicated to the public and between APHIS and FSIS, and controls over the regulatory process were inadequate. USDA's defenders countered that Canada's safeguards are at least equivalent to those of the United States, and that restoring all cross-border trade is critical for the United States to convince other countries that U.S. beef is safe, they asserted.

In Congress

A major issue for Congress has been how best to regain lost markets like Japan and Korea. Among other issues have been whether expanded federal actions are needed to further protect the public and the cattle herd; the validity of the evolving science behind such actions; and the costs and benefits of such actions for consumers, taxpayers, and industry. Besides the safeguards discussed above, Congress has also examined such BSE-related issues as the need for improved labeling and/or traceability of livestock and livestock products, and funding for the government's BSE activities.

Japan and Korea

The sluggish progress on Japan and Korea has frustrated the beef industry and many lawmakers, several of whom have proposed retaliatory action. For example, several bills introduced in 2006 would have required trade sanctions against Japan if the market did not soon reopen. In 2005, a Senate floor amendment to the FY2006 USDA appropriation (H.R. 2744), which would have blocked a new U.S. rule to permit some Japanese beef imports unless Japan lifted its own ban, was deleted from the final conference agreement (H.Rept. 109-255, P.L. 109-97). A number of lawmakers in late 2006 urged a suspension of negotiations with Korea for a bilateral trade agreement until the beef situation was corrected.

COOL

Under the 2002 farm bill, mandatory country of origin labeling (COOL) for fresh beef (among other commodities) in supermarkets was initially slated to take effect on September 30, 2004, but Congress has twice delayed implementation, most recently (under P.L. 109-97) until September 30, 2008. In the 109th Congress, bills were offered to replace the mandatory program for meats with a voluntary program, or conversely, to reinstate an earlier implementation date for mandatory COOL.

Animal Identification (ID)

Some Members have complained that lack of a nationwide animal ID system has hindered investigations into the U.S. BSE cases. USDA, state animal health authorities, and industry groups have been working to create such a program for several years, but progress has been slowed by differences over such issues as whether it should be publicly or privately run; if, and how, to protect the privacy of producer records; and who should pay. In the 109th Congress, bills were offered to require USDA to establish a nationwide electronic animal identification system; to protect the information provided by producers from unauthorized scrutiny and use; and to create a "Livestock Identification Board" with voting members from industry to oversee a national program.

BSE Funding

Total USDA spending for BSE in FY2005 was estimated at $123 million, of which $69 million was for BSE testing (and most of that for the special surveillance program noted above), $49 million to launch the animal ID effort, and $3 million for research. For FY2006, the Administration requested $66 million for USDA's BSE-related activities, including $33 million to continue work on an animal ID program, $21 million for BSE testing/surveillance, and $12 million for research. FDA's BSE request for FY2006 was for nearly $30 million. The FY2006 USDA appropriation (P.L. 109-97) generally covered these requests.

For More Information

CRS Reports

- CRS Report RL32199, Bovine Spongiform Encephalopathy (BSE, or 'Mad Cow Disease'): Current and Proposed Safeguards
- CRS Report RL32932, Bovine Spongiform Encephalopathy (BSE, or 'Mad Cow Disease') in North America: A Chronology of Selected Events
- CRS Report RS21709, Mad Cow Disease and U.S. Beef Trade
- CRS Report RL32012, Animal Identification and Meat Traceability
- CRS Report 97-508, Country-of-Origin Labeling for Foods

Selected BSE Websites

- APHIS: [http://www.aphis.usda.gov/newsroom/hot_issues/bse.shtml]
- FSIS:[http://www.fsis.usda.gov/Fact_Sheets/Bovine_Spongiform_Encephalopathy_BSE/index.asp]
- FDA: [http://www.fda.gov/oc/opacom/hottopics/bse.html]
- Canada: [http://www.inspection.gc.ca/english/anima/heasan/ disemala/ bseesb/bs eesbindexe.shtml]
- OIE: [http://www.oie.int/eng/info/en_esb.htm]

REFERENCES

[1] This report, which replaces CRS Issue Brief IB10127, *Mad Cow Disease: Agricultural Issues for Congress*, summarizes and updates information in other CRS reports, listed on page 6. Sources for facts and citation to reports and studies can be found in these CRS reports.

In: Food Safety and Nutrition ISBN: 978-1-60456-335-1
Editors: J. Miller, P. Bonete, pp. 105-112

Chapter 7

FOOD LABELING: ALLERGY INFORMATION*

Donna V. Porter
Specialist in Life Sciences Domestic Social Policy Division

ABSTRACT

Media attention to food allergies is the result of the recent tracking of food allergy sufferers and a clear rise in the number of affected individuals. Several efforts are underway to improve the ability of individuals who have a food allergy to avoid products that cause symptoms that can range from mild to serious. The Food and Drug Administration (FDA) and the Food Allergy Issues Alliance each have released guidelines to address the issues of labeling and cross-contamination. Nine state attorneys general have petitioned FDA for stricter rules, which are also supported by some consumer groups. The FY2002 agriculture appropriations bill directed FDA to address and report on cross-contamination; however, this report had not yet been submitted. Most recently, the Food Allergen Labeling and Consumer Protection Act of 2004 (P.L. 108-282) was enacted on August 2, 2004. This report will provide background on food allergies and review efforts to provide improved labeling information for food allergy sufferers; it will be updated to reflect legislative or other activity.

* Excerpted from CRS Report RS20975, dated September 19, 2005.

Background

The media attention to the reported incidence of food allergies and a number of fatal reactions seems to be the result of recent government tracking of the incidence of this health problem as well as a recognized increase in the number of cases.[1] The most common food allergies are caused by milk, tree nuts, fish, shellfish, soy, eggs, peanuts and wheat. These foods account for about 90% of all food allergies. Food allergies are estimated to affect about 4 million individuals in the United States. Sensitive individuals can develop a range of symptoms that can be mild or life-threatening allergic reactions, if exposed to a causative allergen. There is no known cure for food allergies and they can only be managed by avoiding foods containing the known allergens. Recent research suggests that a monoclonal antibody being studied may provide short-term protection against most unintended ingestion of peanuts.[2]

A number of different conditions tend to fall under the heading of food allergies in the common vernacular. In most cases, the term "allergy" is used to describe the reaction to a food ingredient that causes a sensitivity response in the person who consumed it. A classic allergy occurs when the body's immune system mistakes a harmless food protein (amino acid sequence) for a virus or bacterium. Antibodies attach themselves to the food molecule which then signal other immune cells to attack by releasing histamines or other destructive chemicals. The body's attempt to rid itself of this "foreign substance" can produce various symptoms, including skin rash, nausea, vomiting, diarrhea, intestinal cramps and spasm, headache or migraine, and swelling in various parts of the body.

A small percentage of food-allergic individuals experience severe reactions, called anaphylaxis. Anaphylactic shock is a relatively rare, but severe and potentially life-threatening, allergic reaction resulting in swelling of the throat and air passage constrictions. Symptoms usually appear within minutes of exposure to the allergen and immediate medical attention is necessary. This reaction requires the intravenous administration of epinephrine (adrenaline) to open the airways, and close medical followup. An estimated 150 Americans die each year from severe allergic reactions to food, such as the reaction to peanuts.

There are several adverse reactions to foods that involve the body's metabolism, but not the immune system, and therefore, the sensitivity is not a true allergy. Food intolerance occurs when the body is unable to metabolize certain foods because of food poisoning or the inability to digest certain food components, such as lactose (milk sugar), usually due to a lack of a particular enzyme. Food aversion is a dislike and avoidance of a food or foods for purely psychological reasons, usually due to an association of that food with a traumatic event or an eating disorder. Food addiction involves the body craving one particular food and suffering withdrawal symptoms when it is removed from the diet. For general sensitivity, the cause is usually unknown, but chemical additives or pesticide residues may be a factor.

FDA Activities

FDA issued an Allergy Warning Letter in June 1996,[3] which sought to alert food manufacturers to the agency's concerns about the labeling of foods that contain allergenic substances. The letter followed numerous reports to FDA concerning consumers who had experienced adverse reactions after exposure to an allergenic substance in a food that did not declare the allergen in the food labeling. While the Food, Drug and Cosmetic Act (FDCA) requires, in virtually all cases, a complete listing of all ingredients of a food, two narrow exemptions appear to provide the reason for recent allergic incidents. The first exemption provides for spices, flavorings and colorings to be declared collectively, without naming each one present. The second exemption from ingredient declarations concerns incidental additives, such as processing aids, that are present in insignificant amounts and do not provide a technical or functional effect in the finished food. For the allergic individual, knowledge that a specific allergenic spice, flavoring or color or incidental additive, even in insignificant amounts, is present, may be very important in avoiding these foods.

In the spring of 2000, FDA received a petition from attorneys general in nine states requesting that the agency issue new labeling regulations for food allergens.[4] The petition requested that a symbol (such as the letter A in a circle) be created and prominently displayed on the upper right corner of food packages, a toll-free hotline be set up for reliable food ingredient information, food labels be required to specify when allergen ingredients were known to be used, and food industry guidelines be established to prevent cross-contamination. Allergenic ingredients can be inadvertently introduced into a food through "cross-contamination," which is the migration of such substances from equipment and utensils resulting in an allergenic food. FDA has yet to respond to this petition in the form of regulations.

In April 2001, FDA released its most recent voluntary guidance on this issue, which is the compliance policy guide entitled Statement of Policy for Labeling and Preventing Cross-contact of Common Food Allergens.[5] The FDA guide recommends complete listing of direct addition of an allergenic ingredient (or sub-ingredient), no ingredient labeling exemptions for incidental additives known to be allergens, and sanitary practices to prevent potential allergen cross-contact. The compliance policy guidance also contains regulatory action criteria for determining when product seizure and legal action are necessary, because of mislabeling or adulteration of the product containing a known food allergen.

FDA announced plans to conduct inspections of plants to assure that allergenic ingredients are not becoming part of the food and candy inadvertently through cross-contamination.[6] The agency stated in 2001 that it would inspect thousands of candy makers, bakeries and other food manufacturers over the next two years. Candy is a particular problem because of the frequent use of nuts and peanut oil in some products and not in others, but cross-contamination occurs due to residues left after inadequate cleaning of assembly line equipment.

The agency convened a public meeting on August 13, 2001 to discuss the labeling of foods containing allergens and the inadvertent addition of allergens due to processing practices.[7] The session focused on whether the source of the ingredient or plain English labeling should be used, the need for supplemental labeling (such as "may contain"), and labeling of ingredients exempted from declaration, such as common or usual names of flavorings, colors and spices. A subsequent workshop on food labeling and allergen declaration was convened in January 2002 in Kansas City at the request of the State of Missouri to address these issues and the particular impact on small business and startups.[8]

In July 2004 FDA announced plans to propose rules by the end of 2004 to require companies to clearly label the color additives carmine and cochineal extract when used in food products, because they can illicit the same symptoms as the other known food allergens. On August 19, 2005, FDA convened a public meeting to obtain comments on defining and permitting voluntary use on food labeling of the term 'gluten-free'.[9] The meeting focused on food manufacturing, analytical methods and consumer issues related to reduced levels of gluten in food. No further regulatory action has been taken on either issue.

Voluntary Guidelines

In May 2001 the Food Allergy Issues Alliance submitted to FDA a consensus document entitled Food Allergen Labeling Guidelines.[10] The Alliance is a group of food trade associations and other interested organizations that convene to discuss allergy issues. The Guidelines are intended to assist food manufacturers to inform food allergic consumers about the presence of the eight major food allergens in their products in consumer-friendly terms. These terms would be listed on the information panel of food products separate from the ingredient declaration. The Alliance seeks FDA clarification through appropriate guidance that presentation of this information would not be considered mislabeling. The Guidelines are presented as an option that would not require new regulations.

The Food Allergen Labeling Guidelines contain four parts. They identify the eight types of allergens that are estimated to cause more than 90% of all food reactions (see listing on page 1). The Guidelines specify the use of ingredient terms commonly understood by consumers so that the words eggs, fish, milk, peanuts, shrimp, soy, walnut or wheat would be disclosed following the label declaration. Product labels complying with these Guidelines would contain a label declaration of a major food allergen, by highlighting its presence in one of several ways (specific statement, asterisk, parenthetical, naming or bolding). Supplemental allergen statements would be used in rare cases where the inadvertent and uncontrollable contact with a major food allergen occurs, although the Guidelines indicate that these statements are not a substitute for good manufacturing practices.

LEGISLATION

In recent Congresses, constituent mail has generated concern about the food allergy labeling issue. Several transportation funding bills have contained provisions that would limit the availability of peanut snacks on airplane flights. However, none of these provisions have ever passed in the final appropriations bills. Several resolutions have been introduced in past years that would direct the Department of Transportation to rescind an earlier DOT directive to establish peanut-free zones on airplane flights when the problem of allergic reactions to air travel peanut snacks first began to occur.

On June 13, 2001, the House Committee on Appropriations adopted an amendment encouraging FDA to promulgate regulations to prevent cross-contamination of foods by undeclared allergens. The Committee adopted the amendment offered by Representative Lowey during its consideration of the FY2002 appropriations bill for the Department of Agriculture and FDA (P.L. 107-76). The amendment also required FDA to report to the Committee by March 1, 2002 on its plans for preventing cross-contamination of foods by undeclared allergens. However, FDA's two-year inspection survey of manufacturing operations, which started in 2001, was to provide the data for this report, which the agency anticipated delivering by the end of 2003. Thus far, this report has not been submitted.

Bills were introduced in the last three Congresses to require the labeling of food products to provide specific information on known allergens. Most recently, the Food Allergen Labeling and Consumer Protection Act of 2004 was introduced as companion bills, S. 741 and H.R. 3684, by Senator Sessions and Representative Lowey and enacted as P.L.108-282 on August 2, 2004. The new law requires that food labeling list in common language any of the eight known food allergens and their food source contained in a product. The labeling requirements apply to raw agricultural commodities, spices, flavorings, colorings, and incidental additives, if food allergens are present. The law also specifies two format options for allergen information to appear on products, and provides for civil penalties for noncompliance after the effective date of January 1, 2006. The legislation requires the Secretary to issue a report to Congress within 18 months of enactment about food allergen cross-contact, advisory labeling and food allergen inspections conducted under this authority. The Secretary is to issue a proposed rule, within two years and a final rule within four years, that defines the term gluten-free for voluntary use in food labeling. The Centers for Disease Control and Prevention will be required to establish a system for tracking food-allergic-related deaths and other clinically significant and serious adverse events. The National Institutes of Health is directed to convene an expert panel to develop recommendations for research activities concerning food allergens within a year. The Secretary is directed to pursue revision of the Food Code to provide guidelines for preparing allergen-free foods in food establishments and provide technical assistance relating to emergency treatment of allergic responses to foods.

Observations on Food Allergy Labeling

For food allergy sufferers, providing ingredient information that will facilitate their selection of foods that do not contain allergens will be useful. Easy identification, by use of a well-recognized symbol and/or specific terms or phrases on packages, frequently suggested in the past, is now mandated under the new law. The requirements for full listing of all known allergenic ingredients, without exceptions for spices, flavoring and colors, the adoption of good manufacturing practices that eliminate the cross-contamination of food products, and identification of incidental additives are several mechanisms that had been sought by various proponents of food allergy labeling. Also suggested have been the elimination of peanut snacks on airlines and peanut butter in school/child care settings. The recent tracking of the incidence and severity of food allergies in some individuals led to increased support for additional labeling information on food packages, since the number of affected individuals seems to have increased.

However, alternative options and additional issues may be raised. Consumer education on the terms, technical or not, and symbols that assist in identifying ingredients that individuals wish to avoid may be necessary to alert food allergy sufferers of the presence of potential allergens. The separate listing of potential allergens, while easily identifiable, may create space and printing size problems on small food packages. Many food allergy reactions occur in situations outside of individuals' homes where label information is not available. The law has addressed this situation by requiring revision of the food code for restaurants, schools, day care, summer camps, grocery store delicatessens and bakeries. In the early 1990s, when sulfites in salad bars were a problem, restaurants used posted signs to alert consumers. Some child care centers and schools have considered an outright ban on peanut butter in their facilities. While this solution appears easy on the surface, it assumes 100% compliance and could create a more hazardous situation, if no one is prepared and trained to assist a child when someone unwittingly sends a peanut butter sandwich to school. Someone could be designated as the individual in a particular school to be prepared to administer injections of epinephrine to a child who has a reaction. However, this decision would be made within local school districts, along with consideration of training and liability.

A recent concern related to food allergy has been development of genetically modified foods and the potential for certain genes to cause food allergies when inserted into other plants and animals. There has been no evidence to date that this technology has resulted in any allergenic problems. Use of these foods will continue to be closely monitored by the federal agencies and outside groups with interest in food allergy problems. FDA's policy continues to be that the presence of any known allergen needs to be labeled. On the other hand, genetically modified foods may hold the potential to eliminate food allergies by removing the allergenic gene from plants or animal products that otherwise need to be avoided by individuals who are allergic to a particular food.

Congressional interest in this issue is likely to continue, with the reports required to be prepared for its use by FDA on regulating the labeling of food allergens, CDC data collection on the food-related allergic response and NIH on research needs. Completion of the reports may become the issue, as the FDA cross-contamination report requested in the FY2002 agriculture appropriation by March 1, 2002 has not yet been transmitted. Future attention will likely be through oversight of the new law's impact on addressing the problems of consumers with allergic reactions to certain food constituents.

REFERENCES

[1] S. Dominus, *The Allergy Prison* New York Times Magazine, Section 6, June 10, 2001, p. 62.

[2] D. Leung, et al., Effect of Anti-IgE Therapy in Patients with Peanut Allergy, *New England Journal of Medicine*, vol. 348, no. 11, Mar. 13, 2003, pp. 986-993.

[3] U.S. Food and Drug Administration, Center for Food Safety and Applied Nutrition, FDA Allergy Warning Letter: Label Declaration of Allergenic Substances in Foods, from Fred Shank, June 10, 1996, p. 3.

[4] Citizen Petition to the Food and Drug Administration, at [http://www.oag.state.ny.us/press/ 2000/may/may26a_attach_00.html].

[5] Food and Drug Administration, Compliance Policy Guide, Compliance Policy Guidance for FDA Staff, Notice to Manufacturers on Policy for Labeling and Preventing Cross-Contact of Common Allergens, at [http://www.fda.gov/ora/compliance_ref/cpg/cpgfod/cpg555-250.htm].

[6] FDA to continue allergen inspections, Food Chemical News, vol. 43, no. 18, June 18, 2001, pp. 27-28.

[7] U.S. Food and Drug Administration, Food Safety and Food Labeling: Presence and Labeling of Allergens in Foods, Announcement of Public Meeting, *Federal Register,* vol. 66, no. 143, July 25, 2001, p. 38572.

[8] U.S. Food and Drug Administration, *FDA Food Labeling and Allergen Declaration*; Notice of Public Workshop, *Federal Register,* vol. 66, no. 246, Dec. 21, 2001, p. 65976.

[9] U.S. Food and Drug Administration, *Food Labeling*; *Gluten-Free Labeling of Foods*; Public Meeting; Request for Comments, *Federal Register,* vol. 70, no. 137, July 19, 2005, p. 41356.

[10] *Food Allergy Issues Alliance*, Food Allergen Labeling Guidelines, May 22, 2001, p. 6.

INDEX

D

E

F

G

H

I

J

K

L

M

N

O

P

Q

R

S

T

U

V

W